A Glance at Food Processing Applications

Edited by Işıl Var and Sinan Uzunlu

Published in London, United Kingdom

IntechOpen

Supporting open minds since 2005

A Glance at Food Processing Applications
http://dx.doi.org/10.5772/intechopen.95722
Edited by Işıl Var and Sinan Uzunlu

Contributors
William Miranda-Zamora, Amirpasha Tirado-Kulieva, David Ricse, Heba Younis, Hassan Abdellatif, Guohua Zhao, Roy Orain Porter, Gözde Ekici, Nadide Gizem Tarakçı, Emek Dümen, Hayato Masuda, Ahmed Albandary, Ayman Hafiz Eissa, Elio Romano, Lajos Helyes, Ayman Ibrahim, Fatemah Albandary, Amit Jaiswal, Vinod Kumar Paswan, Urmila Choudhary, Basant Kumar Bhinchhar, Sheela Kharkwal, Satya Prakash Yadav, Prity Singh, Isıl Var, Sinan Uzunlu

First published in London, United Kingdom, 2022 by IntechOpen
IntechOpen is the global imprint of INTECHOPEN LIMITED, registered in England and Wales, registration number: 11086078, 5 Princes Gate Court, London, SW7 2QJ, United Kingdom
Printed in Croatia

British Library Cataloguing-in-Publication Data
A catalogue record for this book is available from the British Library

Additional hard and PDF copies can be obtained from orders@intechopen.com

A Glance at Food Processing Applications
Edited by Işıl Var and Sinan Uzunlu
p. cm.
Print ISBN 978-1-83969-767-8
Online ISBN 978-1-83969-768-5
eBook (PDF) ISBN 978-1-83969-769-2

For EU product safety concerns:
IN TECH d.o.o., Prolaz Marije Krucifikse Kozulić 3, 51000 Rijeka, Croatia, info@intechopen.com or visit our website at intechopen.com.

Meet the editors

Dr. Işıl Var is a professor in the Food Engineering Department, Çukurova University, Turkey. She received a BSC from Hacettepe University, an MSc from the Scientific and Technological Research Council of Turkey, and a Ph.D. in Public Health Laboratory, England. She completed postdoctoral studies in Molecular Biology at the University of Sussex, United Kingdom between 1993 and 1995. She has published more than 100 papers in national and international peer-reviewed journals and more than 50 international and 75 national presentations in the field of food microbiology, molecular biology, and mycotoxins. She has received several awards from the Scientific and Technological Research Council of Turkey and the VTT Technical Research Centre of Finland.

Dr. Sinan Uzunlu is an associate professor in the Food Engineering Department, Alanya Alaaddin Keykubat University, Turkey. He received his BSc from Çukurova University, Turkey in 1999. He obtained an MSc from Akdeniz University, Turkey, and a Ph.D. from Çukurova University in 2002 and 2012, respectively. He completed postdoctoral visits at the University of Reading, United Kingdom between 2014 and 2017. He has published fifteen papers and two book chapters in national and international peer-reviewed journals and books in the fields of food microbiology and food packaging. Together with Professor Işıl Var, Dr. Uzunlu filed a national patent on modified atmosphere packaging. He is a member of national and international societies and has received awards from national (YÖK, TÜBİTAK) and international (Society of Chemical Industry, UK) organizations.

Contents

Preface XI

Section 1
Introduction 1

Chapter 1 3
Introductory Chapter: Food Processing
by Sinan Uzunlu and Işil Var

Section 2
Processing 7

Chapter 2 9
Effects of the Incorporation of Arabinoxylans Derived from Selected Cereals (Rice Bran and Corn Fibre) and Sugarcane Bagasse on the Quality of Baked Foods: A Systematic Review
by Roy Orain Porter

Chapter 3 45
Honey Production Process
by Emek Dümen, Nadide Gizem Tarakçı and Gözde Ekici

Chapter 4 55
Chocolate: Health, Processing, and Food Safety
by Ahmed Albandary, Fatemah Albandary and Amit K. Jaiswal

Section 3
Packaging 73

Chapter 5 75
Pectin and Its Applicability in Food Packaging
by Heba Younis, Guohua Zhao and Hassan Abdellatif

Chapter 6 93
Utilization of Agro-Industrial Wastes as Edible Coating and Films for Food Packaging Materials
by Urmila Choudhary, Basant Kumar Bhinchhar, Vinod Kumar Paswan, Sheela Kharkwal, Satya Prakash Yadav and Prity Singh

Section 4
Computational Applications **111**

Chapter 7 **113**
Computational Applications for the Evaluation and Simulation of the Thermal Treatment of Canned Foods
by William Miranda-Zamora, Amirpasha Tirado-Kulieva and David Ricse

Chapter 8 **135**
Enhancement of Heat Transfer Using Taylor Vortices in Thermal Processing for Food Process Intensification
by Hayato Masuda

Chapter 9 **157**
Thoughts for Foods: Imaging Technology Opportunities for Monitoring and Measuring Food Quality
by Ayman Eissa, Lajos Helyes, Elio Romano, Ahmed Albandary and Ayman Ibrahim

Preface

The foods for human consumption are produced from plant-based or animal-based raw materials. Excluding raw foods for consumption, raw materials are required to be processed for the production of industrial foods. These materials can be processed in numerous ways, but there is currently a focus on implementing innovative, environmentally friendly, and cost-effective solutions to ensure good food quality and variety.

Post-harvested raw materials exhibit differences in physical, chemical, and biological compositions. Carbohydrates, proteins, lipids, vitamins, minerals, and water are the main components of foods and these components directly determine the design of food processing techniques. Several properties (sensorial, physical, chemical, and microbial) should be considered when applying food processing technologies to fulfill product requirements.

For the increasing population all over the world, the main problem seems to be hunger and insufficient food. As such, the food industry is attempting to ensure food safety and provide food security. Although new technologies have great potential in the production and supply chain of food, they are open to ethical, philosophical, and religious debate. For example, sustainable and nano-technological foods such as laboratory meat, insect-derived proteins, vegetable proteins, and myco-proteins are only just beginning to be considered. Studies are being carried out to provide food to crews in space travels. In addition, strict regulations, applications, and devices are being developed to combat the problem of food waste. New automated devices are being implemented to monitor, process, and pack food. Global Positioning System (GPS) and blockchain technologies are being implemented for a traceable food production system.

This book presents a comprehensive review of food processing applications. Chapter 1 provides an introduction to food processing. Chapter 2 discusses the usage of rice bran, corn fibre, and sugarcane bagasse and their influences on the quality of baked foods. Chapter 3 discusses honey production processes. Chapter 4 discusses the benefits and negative impacts of consuming chocolate and provides the process of manufacturing the product. Chapter 5 examines the potential usage and application of pectin in food packaging. Chapter 6 highlights the application of agro-industrial wastes for packaging processes. Chapter 7 evaluates computational applications for canned foods. Chapter 8 discusses Taylor vortices in thermal food processes. Finally, Chapter 9 provides information about imaging technologies to monitor food quality.

We would like to thank the authors who shared their works with us. We are also grateful to Author Service Manager Mrs. Jasna Bozic at IntechOpen for her assistance throughout the publication process. We also thank Alanya Alaaddin Keykubat University and Çukurova University for their support.

Dr. Işil Var
Professor,
Faculty of Agriculture,
Department of Food Engineering,
University of Cukurova,
Adana, Turkey

Dr. Sinan Uzunlu
Food Engineering Department,
Engineering Faculty,
Alanya Alaaddin Keykubat University,
Alanya, Turkey

Section 1

Introduction

Chapter 1

Introductory Chapter: Food Processing

Sinan Uzunlu and Işil Var

1. Introduction

Evidence suggests that the earliest food processing was undertaken in human history, using the heating process to make raw foodstuffs more palatable. When agricultural activities are started, the need for storage and preservation of raw food materials arose by 3000–1500 BC. Drying under sunlight, fermenting plant-based foods, grinding cereals, and baking bread in oven were the earliest attempts of food processing techniques. Exchanging of foods in trade and explorers-oriented technologies resulted in the change of food processing techniques in distinct food products (e.g., dairies, bakery, fermented foods). All the efforts brought today's reached point [1].

In today's world, food processing is a part of manufacturing industry. What we consume as foods are produced from plant-based or animal-based raw materials, in the meantime called as agricultural sources. Apart from directly consumed foods (e.g., raw eaten fruits) at postharvest term, the raw materials are required to be processed for a healthy consumption, as intermediate or finished value-added food products. Energy, equipment, labor-ship, science are used for a step (unit operation) or a series of steps (process) to produce a marketable product. For example, exposing milk to a heat source could be given as a simple example for a unit operation in terms of heat treatment (pasteurization or sterilization) in the case of milk processing at dairy industry. Therefore, from starting commercial sterilization the way what we observed up to now shaped the food industry [1, 2].

To meet the needs (e.g., shelf life stabile, nutritive, and variety in convenience with diet) of the global market, postharvested raw materials are processed in different ways. To serve a marketable food product, there are a number of intrinsic and extrinsic parameters that determine specific processing design of each product. Raw materials as sources of foods to be used for processing are complex substances. As intrinsic parameters the content of foods, carbohydrates, proteins, fats, vitamins, minerals along with water are first accounted. However, biological, chemical, and physical properties might be classified both in intrinsic and extrinsic parameters. The content and surrounding atmosphere (e.g., in-pack conditions, storage room conditions) of foods vary for almost each food product. Biological (bacteria, yeast and molds, viruses, parasites), chemical (pesticide, fungicide, allergens, mycotoxins), and physical (stones, dirt, metal, glass, insect fragments, hair) hazards should be controlled for a safe, nutritious, and wholesome consumption.

The basic processing steps consist of raw material harvesting, pretreatment (e.g., washing, separation,), basic unit operations (e.g., heat treatment, freezing, drying), and packaging. There are currently a number of different ways of processing technologies being applied in different ways at industrial scale. It is clear that almost every food types need different processing conditions to be an optimal product in its distinctive package. An overall view to food processing is presented in **Table 1**.

Unit operations in food processing	Thermophysical properties, microbial aspects, other considerations	Common food preservation/ processing technologies	Other food processing/ preservation technologies
Heat transfer	Raw material handling	Processes using addition or removal of heat (pasteurization and blanching, thermal sterilization, aseptic processing, sous-vide cooking, microwave heating, ohmic heating, drying, refrigeration and freezing)	Fermentation
Fluid flow	Cleaning and sanitation	Nonthermal food processing and preservation (irradiation, high-pressure processing, pulsed electric field processing, ultrasound)	Extrusion
Mass transfer	Engineering properties of food, biological, and packaging material		Baking
Mixing	Microbiological considerations		Hurdle technology
Size adjustment	Role of acidity and water activity in food safety and quality		Packaging
Separation	Reaction kinetics		

Table 1.
An overview to food processing. Adapted from Park et al. [2].

However, both thermal and nonthermal food processing technologies are up-to-date widely used, alternative food processing technologies are now involved in the field. Some examples for these thermal technologies, microwave, radiofrequency, infrared heating, pressure-assisted thermal sterilization, and sous-vide processing, could be given. Whereas, nonthermal technologies in generally apply high hydrostatic pressure, irradiation, ultrasound, pulsed electric field, pulsed light technologies, and 3D printing. Nanotechnology (e.g., bio-sensing, packaging, agro-chemical production) is being started to implement besides these traditional processing techniques. The novel nonthermal or cold-pasteurization technologies provide less energy and water usage than traditional processing technologies, to sustain the scarce resources in globally. These technologies in sum serve for better quality, more healthful, minimally processed, traceable, and safer foods. Beyond product performance, the expectation of consumers is to consume safe foods that are produced under hygienic and sanitary conditions [3–5].

The food industry ensures food safety with the rise of pasteurization and sterilization techniques and food security with modern agricultural practices at reasonable prices. Meanwhile, the main problem still seems to be the issue of hunger and insufficient food, and the demand for food for the growing world population is still under discussion. Moreover, to meet the dynamic conditions and consumer-oriented needs, synchronized innovative solutions in food processing technologies should be found and applied.

So far all the time, the urgent need was to feed the world, by growing sufficient raw material and producing food products (e.g., wheat flour, dairies, meat products, aqua culture products, etc.) at reasonable prices. However, in today's world, the climate crisis threatens sustainable food production, whereas sustainability concern has now become a real endangering issue. The United Nations defined "*the needs of the present without compromising the ability of future generations to meet their own needs*" for sustainable development [6]. From starting this end, a food production should ensure secure, environmentally sustainable, and healthy supply. The food industry produces more accessible and convenient foods, in terms of having reasonable price and remaining stable in predetermined shelf life periods. However, the quality of the foods, mainly snacks or on-the-go food or beverages, should be questioned for quality aspects. These types of foods and beverages containing carbohydrates, lipids, sugar, and food additives, which are increasing day by day and are purchased relatively cheaply, should be discussed for their unhealthy conditions. For the case of this concern in food industry, a high amount of negative perception on processed foods has become among consumers. The rate of literate people, social media, TV programs, magazines have all contributed to aforementioned negative concern on foods in pack.

To overcome current negative conditions, efficient and economic use of scarce resources (water, land, energy, air), environmental protection, and waste (food and non-food) management are requiring innovative attempts, unless carrying no suspicion on health concern of consumers.

Author details

Sinan Uzunlu[1]* and Işil Var[2]

1 Food Engineering Department, Engineering Faculty, Alanya Alaaddin Keykubat University, Alanya, Turkey

2 Faculty of Agriculture, Department of Food Engineering, University of Cukurova, Adana, Turkey

*Address all correspondence to: suzunlu@hotmail.com

References

[1] Fellows PJ. Food Processing Technology. Cambridge, UK: Woodhead Publishing; 2009

[2] Park SH, Buddhi PL, Balasubramaniam VM. Principles of food processing. In: Clark S, Jung S, Lamsal B, editors. Food Processing: Principles and Applications. West Sussex, UK: John Wiley & Sons Publishing; 2014. pp. 1-15

[3] Neetoo H, Chen H. Alternative food processing technologies. In: Clark S, Jung S, Lamsal B, editors. Food Processing: Principles and Applications. West Sussex, UK: John Wiley & Sons Publishing; 2014. pp. 137-169

[4] Gunasekaran S. Nanotechnology for food: Principles and selected applications. In: Clark S, Jung S, Lamsal B, editors. Food Processing: Principles and Applications. West Sussex, UK: John Wiley & Sons Publishing; 2014. pp. 171-205

[5] Schoenfuss T, Lillemo JH. Food safety and quality assurance. In: Clark S, Jung S, Lamsal B, editors. Food Processing: Principles and Applications. West Sussex, UK: John Wiley & Sons Publishing; 2014. pp. 233-247

[6] Murphy F, McDonnell K, Fagan CC. Sustainability and environmental issues in food processing. In: Clark S, Jung S, Lamsal B, editors. Food Processing: Principles and Applications. West Sussex, UK: John Wiley & Sons Publishing; 2014. pp. 207-232

Section 2

Processing

Chapter 2

Effects of the Incorporation of Arabinoxylans Derived from Selected Cereals (Rice Bran and Corn Fibre) and Sugarcane Bagasse on the Quality of Baked Foods: A Systematic Review

Roy Orain Porter

Abstract

The supplementation of baked foods, namely cookie/biscuits, bread and cakes with agricultural by-products from cereal based fibres (rice bran and corn fibre) and sugarcane bagasse at rates of 0% - 15%; 0% - 30% and 0% - 10% respectively can significantly improve its nutritive value and enhanced its physical and sensorial qualities. This chapter aims to review the role of dietary fibres derived from selected cereals (rice bran and corn fibre) and sugarcane bagasse in baked foods, namely cookies/biscuits, bread and cakes; evaluate their effects on the physical and sensory qualities of these baked food products and to critically assess their beneficial impacts in baked foods. These enriched food products can potentially be utilised in shaping health policies, contribute to the dietary fibre needs of consumers and facilitate the development of functional foods. Fibre enriched foods potentially can assist in improving various physiological functions of the human body. A Keyword-based search strategy was utilised to conduct a comprehensive search for articles catalogued in ScienceDirect, Web of Science, PubMed, Medline, CINAHL and Google Scholar that were published between January 1, 2010 and August 1, 2020. Applicable aspects of the Preferred Reporting Items for Systematic Reviews and Meta-Analyses (PRISMA) guidelines provided the framework of this review. Fourteen (14) studies met the inclusion/extraction criteria and was placed into sub-groups by food types and fibre used in supplementation. Only eleven (11) studies were suitable for statistical data analysis. The supplementation of sugarcane bagasse at both 5% and 10% and rice bran up to 15% into cookies/biscuits significantly undesirable acceptance ($p < 0.05$). Corn fibre enriched cookies/biscuits up to 20% showed a significantly ($p < 0.05$) favourable impact on the sensory qualities of the food product. The physical qualities of sugarcane bagasse supplemented cookies/biscuits were negatively affected. The incremental addition of sugarcane bagasse resulted in at 50% rise in the firmness of 10% enriched cookies/biscuits, from 5.7 $\pm$ 5.4 (Kg Force) to 13.0 $\pm$ 3.9 (Kg Force). Corn fibre cookies supplementation did not significantly affect its physical qualities. Rice bran incorporation of 15% in bread showed a significant ($p < 0.05$) undesirable effect on its sensory qualities. However, the was no significant adverse effect on its physical quality. Corn bran

IntechOpen

enriched cakes up to 20% fibre incorporation displayed a significant ($p < 0.05$) favourable effect on the sensory properties of cakes.

Keywords: Arabinoxylans incorporation, dietary fibres, rice bran, corn fibre, sugarcane bagasse, baked food products

1. Introduction

There are numerous dietary fibre enriched food products developed in the food industry during the last decade encompassing various popular and widely consumed foods such as bread, cakes, cookies or biscuits, yoghurts among others [1–4]. It is well established in the literature, that dietary fibre intake at levels greater than 25 g per day, tend to be associated with numerous health benefits, namely the reduced risk of coronary heart disease, type 2 diabetes, enhanced physiological functions of the human body, improved weight maintenance and other positive effects on various disease risk factors and the alleviation of certain types of cancers [5–8].

In recent years, several drivers such as consumer awareness of the nutritional value of dietary enriched foods, and governmental policies promoting healthy lifestyle behaviours have contributed to the continual increase in the use of dietary fibres in foods [3, 9–11]. Consequently, the value of the dietary fibres global market is expected to experience an astonishing annual rise, with the latest estimates projected growth of about 9.74 billion U.S. dollars by 2025 [12]. Dietary fibres can be considered as non-digestible carbohydrates which are inclusive of lignin, resistant oligosaccharides, resistant starch, non-starch polysaccharides (NSP) such as cellulose, pectins, hydrocolloids and hemicelluloses of which can be eaten and are not prone to enzymatic digestion and absorption within the small intestines, but can undergo complete or partial fermentation in the large intestine of the human body [1, 3, 6]. In previous studies, Foschia, Peressini, Sensidoni, and Brennan [13] indicated that the major dietary fibres (DFs) are consist of arabinoxylans, β-glucans, resistant starch and inulin. In the non-starch polysaccharides (NSP) fraction, arabinoxylan is the main polysaccharide, additionally, arabinoxylan structure is comprised of a framework of β-(1–4) connected xylose residues to which α-L-arabinose tend to linked unto the second or third carbon positions [14, 15]. Agricultural by-products from milling industries namely fibres from selected cereal crops rice, corn, and energy crop sugarcane and plant parts from other fruits and vegetables can be regarded as dietary fibres and subsequently tend to contain arabinoxylans in varying amounts [16–18]. The addition of fibres to food products tends to influence the consistency, texture, rheological tendencies and sensory characteristics of the finished food products [3]. The incorporation of fibres in breakfast cereals, bread, cookies, cakes, pasta, yogurt, beverages and meat products among others have been widely reported with desirable results [2, 13, 19–23].

Arabinoxylans have been considered as an important dietary fibre of selected cereals (rice bran and corn fibre) and sugarcane bagasse and it has been suggested that when incorporated in the optimum proportions into food products they are capable of improving its quality, which includes, but not limited to only changes in the physical, rheological, and sensorial characteristics of food products [24–26]. Moreover, various technological functions of food products are also enhanced by dietary fibre incorporation into foods such as its nutritional value, functional properties and improvement of other chemical properties [27–29]. It has also been reported in several studies that dietary fibre arabinoxylans from agricultural by-products such as rice bran, corn fibre: brans and other corn parts and sugarcane bagasse can be safely utilised in the baking industry and consequently be used whole or as extracts of soluble dietary arabinoxylans to facilitate the production of functional and health-promoting food products through the supplementation of bread, cakes, cookies and other food products at varying incorporation rates [2, 17, 30–34].

However, there is a lack of consensus in the body of literature relating to the beneficial effects the incorporation at various levels of dietary arabinoxylans derived or originating from sources such as selected rice bran, corn fibre (bran and other parts) and sugarcane bagasse have on the sensory and physical qualities of baked foods including cookies/biscuits, bread and cakes [23, 35–40]. This systematic review, therefore, endeavours firstly to review the role of dietary fibre derived from selected cereals (rice bran and corn fibre) and sugarcane bagasse in baked foods, namely cookies/biscuits, bread and cakes; secondly to evaluate the effects of the incorporation of dietary fibre derived from selected cereals (rice bran and corn fibre) and sugarcane bagasse on the physical and sensory qualities of these baked food products and finally to critically assess the beneficial impacts of dietary fibre incorporation derived from selected cereals (rice bran and corn fibre) and sugarcane bagasse in baked foods, including bread, cakes and cookies.

2. Methodology

The methodology follows applicable aspects of the Preferred Reporting Items for Systematic Reviews and Meta-Analyses (PRISMA) guidelines [41] and includes details of the search strategy, studies selection, inclusion and extraction criteria, data extraction and assessment of study validity, risk of bias assessment and data analysis of the relevant studies utilised in this systematic review.

2.1 Research strategy

A systematic literature search using a Keyword-based concept was conducted for articles catalogued in ScienceDirect, Web of Science, PubMed, Medline, CINAHL and Google Scholar that were published between January 1, 2010 and August 1, 2020 using the following search strategy. The keyword-based searches included Boolean operators and were constructed using words from the research question along with the combination of truncations and wildcards to access as much primary research material as possible. The searches were limited to Scholarly and Peer-reviewed and English language. Grey areas of the literature such as government reports, conference reports and food magazines were also searched. Further, the reference lists of recent systematic reviews were searched for additional references.

2.2 Study selection, inclusion and exclusion criteria

The eligibility criteria utilised for the selection of relevant research studies followed the population, intervention, comparison, outcome and study settings (PICOS) research question framework. Further details are summarised in **Table 1**.

For studies to be included in the systematic review the following criteria were established, namely:

- The keywords and phrases e.g. arabinoxylans incorporation, rice and corn brans/fibre, sugarcane bagasse, baked food products among others relating to the research question should be included in the particular article title,

Population	Human subjects and quantitative parameters of nutritional and physical properties of baked foods including bread, cakes and cookies.
Intervention	Rice bran, corn fibre and sugarcane bagasse
Comparison	Nutrient profile, volume, texture, weight, height, colour and sensory evaluation metrics.
Outcome	Improve quality
Study Setting	All research studies investigating the effects of the incorporation of dietary fibres on the nutritional, physical and sensory qualities of baked food products.

Table 1.
PICOS criteria for study selection.

- Human subjects including adults/children,
- Food vehicles included bread, cakes and cookies,
- Source of arabinoxylans rice bran, corn bran and sugarcane bagasse,
- Quantity of arabinoxylans from the particular sources that was incorporated,
- Sensory evaluation,
- Quantitative parameters of nutritional profile e.g., fibre, fat, carbohydrates and protein content and physical profile e.g., texture, volume, weight, height and colour of baked foods including bread, cakes and cookies,
- Full text of articles (abstract only of articles would not be considered),
- Articles should be written in English language,
- The published dates of articles should be between January 1, 2010 and August 1, 2020,
- Controlling bias by using randomisation and mean measurements of food products from triplicates instead of a single measurement and
- Articles should be able to be placed into sub-groups to facilitate statistical analysis.

Studies under consideration for inclusion in the review which did not contain the required information as outlined previously were not selected. All articles chosen were sent to the EndNote reference database, which facilitated the identification of duplicate articles, which were also excluded from the review.

Articles consisted of a sensory evaluation of a particular baked food product, including bread, cakes and cookies; a nutritional and physical profiles of bread, cakes and cookies were deemed fundamental to the review since they provided data relating to the nutritional, physical and sensory qualities of baked foods being studied. The use of articles investigating the different dietary fibre sources of arabinoxylans (rice and corn brans and sugarcane bagasse) incorporation into foods facilitated the assessment of the effects of arabinoxylans incorporation into foods, its possible implications, roles and the possible optimum inclusion proportions of arabinoxylans, which may result in improved nutritional, physical and sensory qualities of different food products for consumer consumption. Priority was given to studies examining arabinoxylans derived from agricultural by-products of selected cereals (rice and corn) and sugarcane bagasse. Articles produced earlier than 2010 were excluded, mainly to reflect the recent advancement of public health policy guidelines, modern processing techniques and equipment within the food industry.

2.3 Data extraction and assessment of study validity

The data extracted from the studies were performed independently by the researcher and it included the pertinent characteristics of the studies relating to the research question of the systematic review. The characteristics used for the data extraction sheet consisted of the following headings:

- Authors' name
- Title
- Location of study and funding source
- The objective of the study
- Source of arabinoxylans and amount used
- Food vehicle (bread, cakes and cookies)
- Control of bias – Randomisation usage in design, triplicate measurements, statistical analysis and sensory evaluation
- Sensory evaluation metrics
- The nutrient profile of food product
- Physical parameters of food product (volume, texture, weight, height, colour)
- Statistical analysis
- Results
- Conclusion

The data from studies were checked for errors and any unavailable data was denoted as not determined (ND) in the particular table.

The assessment of study validity was conducted using the Downs and Black checklist [42]. This checklist is comprised of 27 questions and can be utilised in the assessment of the methodological quality of both randomised and non-randomised studies [42]. However, for this review only fourteen (14) questions were found to be applicable, the other questions were denoted as not applicable. The score ranges and corresponding quality levels used for the Downs and Black [42] were as follows: excellent (26–28); good (20–25); fair (15–19) and poor (≤14) [43]. Some of the included criteria relevant in the assessment of articles for this type of review comprised of the following areas:

i. Reporting

ii. External validity

iii. Bias

iv. Confounding factors

v. Statistical power

2.4 Risk of bias assessment

The risk of bias assessment of the articles used in this review was evaluated using the Cochrane collaboration's tool for assessing the risk of bias. The statistical information presented in the articles was assessed for its appropriateness. The outcomes reported in articles used were verified for accuracy and the section of the study it was first reported was noted. Moreover, the credentials and attachments of the respective authors of the articles utilised in the review were also checked. Importantly, references to the disclosure of interest were keenly examined at the end of the articles.

2.5 Data handling

In this review, only eleven (11) research articles out of fourteen (14) studies were considered suitable for statistical data analysis. There were five (5) studies that focused on rice bran as the arabinoxylan source and both cookies/biscuits (3 studies) and bread (2 studies) as the food vehicle; four (4) studies used corn fibre as the arabinoxylan source and both cakes (2 studies) and cookies/biscuits (2 studies) and finally two (2) studies conducted research using sugarcane bagasse as the arabinoxylan source and cookies/biscuits as the food vehicle. In the area of sensory evaluation, various scales were used in the assessment of particular sensory attributes such as hedonic scales (5-point, 7-point and 9-point) and10 cm unstructured line scale. Thus, sensory assessment scores were standardised by dividing individual scores given by the panel by the maxima of the particular scale used, then multiply by 100 to convert to a percentage. Data of panellists were extracted and grouped into categories of trained, semi-trained and untrained; gender and nationality and utilised to conduct a descriptive statistical analysis. Students were grouped as semi-trained and in studies that did not specify the training of panellists, they were classed as untrained. Studies were grouped into sub-groups according to the food vehicle and source of arabinoxylan that was incorporated into the food product. The weighted means of the papers was calculated by using the Statistical Package for Social Sciences (SPSS), version 26 software. Moreover, the percentage of fibre incorporation was also similar for each sub-group utilised for data analysis. The metrics extracted from the included studies relating to the physical analysis of the particular food vehicle were also standardised namely for bread: volume (ml), mass (g) and specific volume (g/ml); cookies/biscuits: width (mm), thickness (mm), spread factor (%) and colour: L*, a* and b* and finally cakes: crumb colour: L*, a* and b*, crust colour: L*, a* and b* and texture (Kg F).

2.6 Data outputs

The fundamental characteristics extracted from the articles were conducted using applicable aspects from the Preferred Reporting Items for Systematic Reviews and Meta-Analyses (PRISMA) guidelines [41]. Statistical analysis of data extracted from the various studies included in the review was performed using a Simple (one-way) analysis of variance (ANOVA) using Statistical Package for Social Sciences (SPSS), version 26. Post Hoc analysis was conducted using Tukey's test to identify where the difference lies between each group, (see Appendix F). The significant difference between the means was identified where ($p \leq 0.05$).

3. Results

3.1 Study selection

In this systematic review, a total of fourteen (14) studies satisfied the inclusion criteria established, the process that guided the selection of these studies is illustrated in the PRISMA diagram (see **Figure 1**). However, only eleven (11) studies facilitated categorisation into a sub-group, namely bread, cakes and cookies/biscuits and then further sub-divided into fibre type and food vehicle and were thus considered suitable to be utilised for statistical analysis, (see **Table 2**). In the fourteen (14) studies included, all conducted a sensory evaluation study, a nutritional analysis/profile and a physical analysis of the particular food produced being studied, except the research studies completed by [17, 23], there was no nutritional analysis/profile and the study carried out by [44], which did not perform a physical analysis of the chapatti, a type of fermented bread. The sensory studies consisted of 362 participants from eleven (11) countries, see **Figure 2**. Panel members used in the

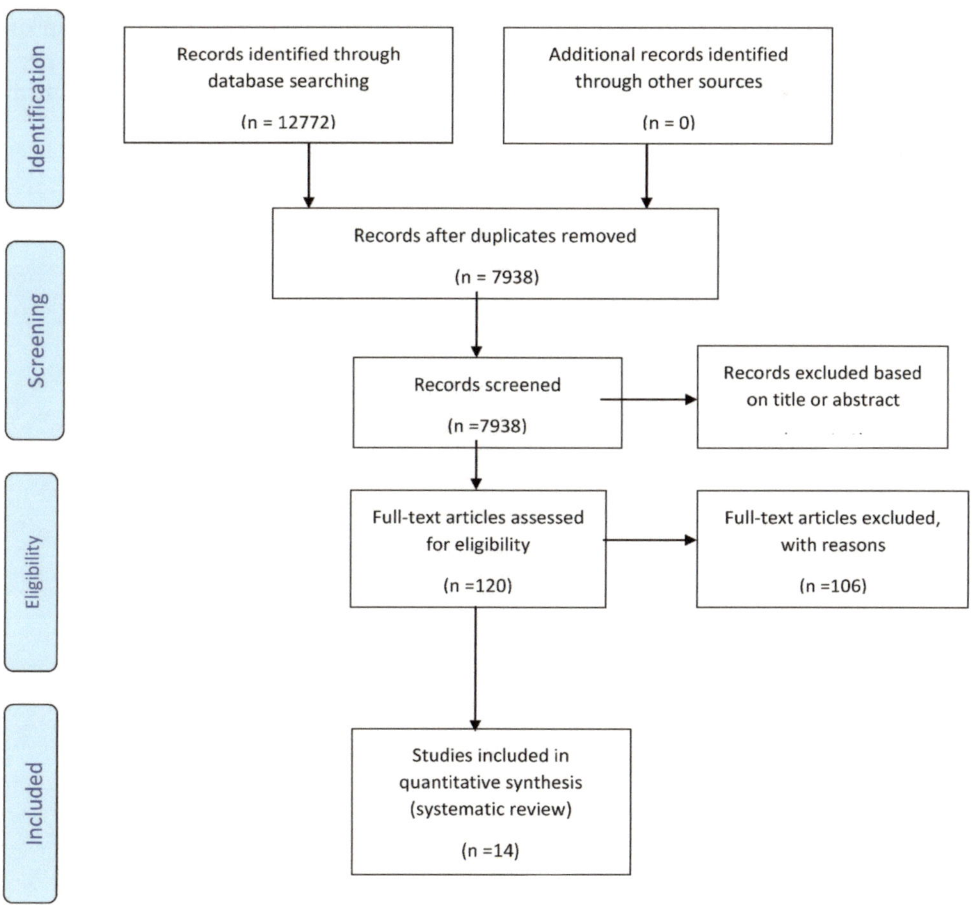

Figure 1.
Study selection process based on PRISMA guidelines. Adapted from [41].

Detail	Appearance	Taste	Texture	Overall acceptance
Control (0%)	85.6 ± 8.6a	86.9 ± 9.8c	86.2 ± 9.2a	86.1 ± 9.2a
5% SCB	84.2 ± 7.5b	82.1 ± 7.4b	82.1 ± 7.4b	84.2 ± 7.5a
10% SCB	68.9 ± 18.9c	69.5 ± 21.0c	72.3 ± 13.4c	72.3 ± 13.4b

Values are means ± standard deviation of sub-groups. Means in the same column with different superscripts are significantly different ($p < 0.05$). Key: SCB – sugarcane bagasse.

Table 2.
Sensory quality of sugarcane bagasse enriched cookies/biscuits.

sensory studies ranged from 5 to 60 participants and the mode were 30 and 10 panellists. The panellists were grouped into trained, semi-trained and untrained, see **Table 2**. The panel members were also made up of both gender (male and female), age groups (which ranged from 18 to 50 years old) and occupations (staff members, post-graduate students and others not mentioned). In addition, eight (8) of the 14 studies were conducted in Asia, namely Sri Lanka - 1, India – 3, Iran – 1, Pakistan – 2 and Bangladesh – 1. Two (2) studies were conducted in North America, namely United States – 1 and Mexico – 1 and two (2) studies were carried out in Africa, namely Cameroon –1 and Nigeria - 1. Finally, one study was conducted in Brazil, South America. Studies examined various baked food products, namely bread - four (4) studies), cakes - three (3) studies, cookies/biscuits studies - seven (7) studies.

The metrics utilised in the sensory studies of this review encompassed colour of crumb and colour of crust; aroma; flavour; appearance; taste; texture firmness or hardness and overall acceptability, see **Table 2**. Meanwhile, for the physicochemical and physical characteristics aspects of the studies, the parameters examined included moisture %, protein %, ash %, fat %, carbohydrates %, crude fibre %,

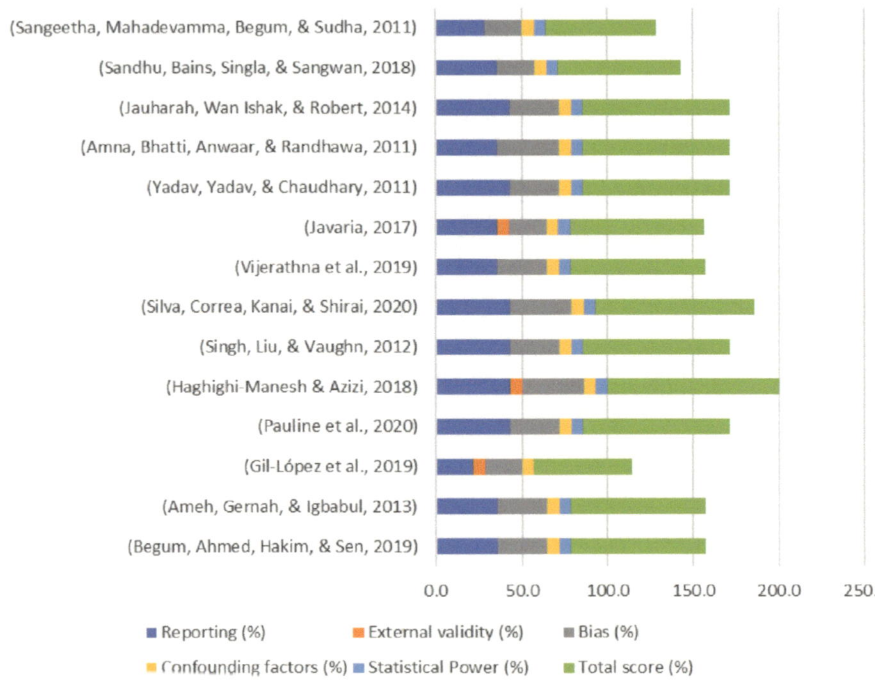

Figure 2.
Quality assessment results of the various studies used in the review, presented in percentages per category and overall total using the Downs and Black checklist.

mineral contents and phenolic contents; texture for example firmness or hardness; colour or luminosity of crust and crumb (for example lightness/brightness, redness/greenness and yellowness/blueness); loaf weight, bread height and bread volume; diameter, thickness, spread ratio and texture, see **Table 2**. The intervention data extracted were the various dietary fibres derived from arabinoxylans selected sources, namely rice bran, corn fibre and sugarcane bagasse and which was supplemented at different rates ranging from 0–30% into baked foods bread, cakes and cookies/biscuits.

3.2 Data quality

The quality assessment of the fourteen (14) research studies included in the review was conducted using the Downs and Black [42] checklists, importantly only fourteen (14) questions were considered applicable, the other questions were denoted as not applicable to this review and consequently were not utilised to assess the included studies. The highest score was received by the research paper [17], namely 14/14, while the lowest score was obtained by the research paper [44], namely 8/14. Five (5) research papers received 11/14; four (4) research papers received 12/14 and one (1) research paper each received scores of 13/14, 8/14, 9/14 and 10/14 respectively, (see **Figure 2**). Thirteen (13) out of the fourteen (14) studies excelled in the category of reporting, only [44] scored poorly. In the category of bias or internal validity all fourteen (14) papers received satisfactory scores, contrasting in the category of external validity only two (2) studies excelled, the other studies failed to show that the panel members were chosen from a representative sample. Moreover, in areas of confounding factors and statistical power, the majority of studies received high scores. Based on the scoring scale of the Downs and Black [42] checklists the quality of the fourteen (14) studies would be considered in the range of fair to excellent [43].

3.3 Data analyses

This section entails the review and analysis of the findings of the primary research articles included in the review which were categorised into sub-groups to reveal common effects and enhance statistical power, except for those papers which did not enable statistical analysis. The studies were grouped as follows: the types of food or food vehicle namely, cookies, biscuits, bread and cakes and fibre; the type of fibre supplemented into the food product, namely rice bran, corn fibre and sugarcane bagasse and the particular outcomes: effects or no effects were outlined or highlighted, (see **Table 2**). The included studies were further sub-divided into specific food and fibre types; the same rates of fibre incorporation into the particular food product as the food vehicle and their respective similar sensory metrics and physical parameters were extracted from the eleven (11) studies to facilitate statistical analysis. The findings will be presented under three (3) main food type headings, namely cookies/biscuits, bread and cakes to facilitate a logical presentation.

3.3.1 Cookies/biscuits: sugarcane enriched

The incorporation of sugarcane bagasse up to the level of 10% showed significantly undesirable overall acceptance ($p < 0.05$). Moreover, based on the sensory evaluation results, as the level of sugarcane bagasse increased, the overall acceptance of enriched cookies/biscuits reduced significantly ($p < 0.05$), see **Table 2**. Cookies/biscuits incorporated with sugarcane bagasse at 5% were similar to control.

Sugarcane bagasse incorporation up to 10% resulted in no significant differences ($p > 0.05$) in the physical quality of the cookies/biscuits. In **Table 3**, the thickness of fibre supplemented cookies was marginally less than the control sample and although the enriched cookies/biscuits were slightly wider, enrichment produced a reduction in the parameter of spread factor and was more 2 times harder than control cookies/biscuits.

3.3.2 Cookies/biscuits: rice bran enriched

Rice bran fibre supplemented cookies/biscuits up to 15% showed significantly undesirable overall acceptance. The sensory scores were all lower than the control sample according to sensory evaluation. Rice bran enriched samples obtained, namely 15% incorporation obtained the low scores for most of the sensory attributes assessed, see **Table 4**. Overall acceptance of cookies/biscuits significantly reduced ($p < 0.05$) in comparison to the control sample of cookies/biscuits. Incorporation of rice bran into cookies/biscuits at 10% performed only slightly better than 5% and 15% levels of incorporation. However, the control sample obtained the best overall acceptance based on sensory evaluation.

The incorporation of rice bran into cookies/biscuits significantly ($p < 0.05$) enhanced the thickness of the food product. The width and degree of spread factor showed no significant differences in comparison to the control. Further details are outlined in **Table 5**.

3.3.3 Cookies/biscuits: corn fibre enriched

Corn fibre supplemented cookies up to 20% obtained a significant ($P < 0.05$) desirable overall acceptance based on sensory panel evaluation. There were significant differences ($p < 0.05$) in colour between enriched corn fibre cookies/biscuits and the control. Interestingly, 20% supplemented obtained the best score for overall

Detail	Thickness (mm)	Width (mm)	Spread factor (SF%)	Texture (Kg force)
Control (0%)	5.3 ± 0.3a	48.9 ± 10.3b	100.0 ± 0.0c	5.7 ± 5.4d
5%	4.9 ± 0.2a	51.0 ± 6.6b	87.3 ± 3.0c	7.6 ± 6.6d
10%	4.9 ± 0.4a	51.9 ± 5.0b	83.4 ± 9.5c	13.0 ± 3.9d

Values are means ± standard deviation of sub-groups. Means in the same column with different superscripts are significantly different ($p < 0.05$). Key: SCB – sugarcane bagasse.

Table 3.
Physical quality of sugarcane bagasse enriched cookies/biscuits.

Detail	Flavour	Colour	Taste	Texture	Overall acceptance
Control (0%)	79.8 ± 14.5a	80.5 ± 15.1a	76.5 ± 13.0a	72.9 ± 11.7a	78.4 ± 12.3a
5% RB	70.9 ± 10.8ab	73.3 ± 11.7ab	69.6 ± 14.2ab	67.3 ± 10.7b	72.1 ± 11.0b
10% RB	72.9 ± 8.5ab	72.4 ± 4.7ab	71.3 ± 7.1ab	71.4 ± 9.7a	72.3 ± 5.4b
15% RB	71.7 ± 5.2ab	71.4 ± 7.2ab	67.4 ± 8.0ab^1	71.5 ± 1.0a	71.0 ± 3.1b

Values are means ± standard deviation of sub-groups. Means in the same column with different superscripts are significantly different ($p < 0.05$). Key. RB – rice bran, ab^1 is significantly different from control and 10% RB samples.

Table 4.
Sensory quality of rice bran enriched cookies/biscuits.

Detail	Thickness (mm)	Width (mm)	Spread factor (SF%)
Control (0%)	8.6 ± 1.0a	42.7 ± 13.6b	51.7 ± 5.2c
5% RB	9.5 ± 0.5ab[1]	40.5 ± 13.2b	49.4 ± 5.6c
10% RB	9.6 ± 0.1ab	38.8 ± 12.9b	46.5 ± 6.6c
15% RB	9.9 ± 0.3ab	36.9 ± 13.8b	44.5 ± 6.9c

Values are means ± standard deviation of sub-groups. Means in the same column with different superscripts are significantly different ($p < 0.05$). Key: RB – rice bran and ab[1] – no significant difference between 5% and 10% RB supplementation.

Table 5.
Physical quality of rice bran enriched cookies/biscuits.

Detail	Flavour	Colour	Appearance	Texture	Overall acceptance
Control (0%)	81.5 ± 10.4a	81.2 ± 10.9a	83.2 ± 10.9a	85.3 ± 6.3a	83.2 ± 10.9a
10% CF	83.8 ± 7.9a	82.6 ± 10.4ab	83.2 ± 10.9a	85.3 ± 6.3a	84.4 ± 8.4ab
20% CF	79.9 ± 13.4ab	79.3 ± 12.8ab	79.9 ± 13.4a	80.8 ± 13.6b	86.7 ± 5.3ab
30% CF	67.8 ± 14.5b	70.0 ± 19.8 ac	67.2 ± 17.7b	67.9 ± 18.2c	67.2 ± 17.7c

Values are means ± standard deviation of sub-groups. Means in the same column with different superscripts are significantly different ($p < 0.05$). Key: CF – corn fibre.

Table 6.
Sensory quality of corn fibre enriched cookies/biscuits.

acceptance. Moreover, both 10% and 20% enriched cookies/biscuits were found to be statistically similar, see **Table 6**.

In **Figure 3**, both 10% and 20% enriched cookies/biscuits obtained the best scores from panellists. Meanwhile, the incorporation of cookies/biscuits at 30% was not well accepted during sensory evaluation.

The physical qualities of up to 30% corn fibre supplemented cookies/biscuits were statistically similar to the control sample. However, cookies/biscuits incorporated with corn fibre up to 20%, showed increased thickness in comparison to the

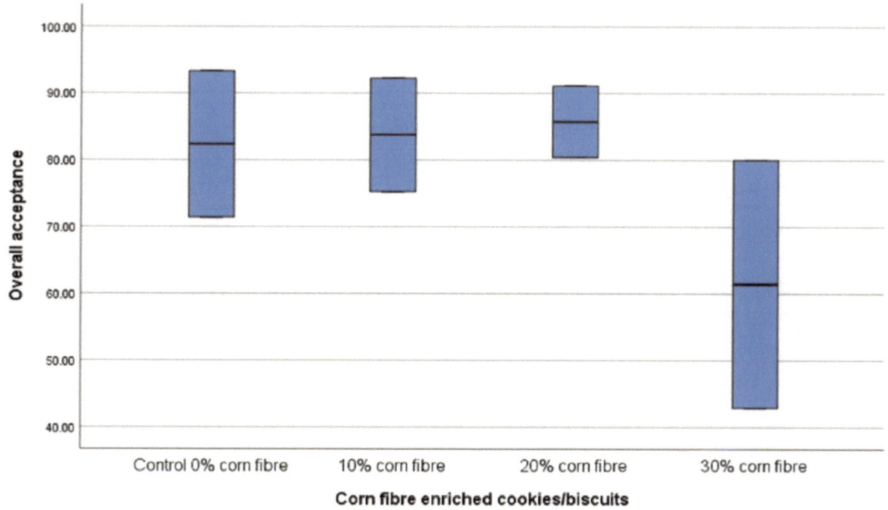

Figure 3.
Graph depicting overall acceptance of corn enriched cookies/biscuits.

Detail	Thickness (mm)	Width (mm)	Spread factor (SF%)	Texture (Kg force)
Control (0%)	5.9 ± 0.4a	49.1 ± 24.5b	76.9 ± 32.7c	1.6 ± 0.7d
10% CF	5.9 ± 0.4a	48.3 ± 19.9b	75.5 ± 25.6c	1.8 ± 0.5d
20% CF	6.1 ± 0.4a	48.4 ± 18.8b	74.0 ± 23.6c	1.9 ± 0.4d
30% CF	5.9 ± 0.7a	51.9 ± 5.0b	75.6 ± 19.5c	2.1 ± 0.2d

Values are means ± standard deviation of sub-groups. Means in the same column with different superscripts are significantly different ($p < 0.05$). Key: CF – corn fibre.

Table 7.
Physical quality of corn fibre enriched cookies/biscuits.

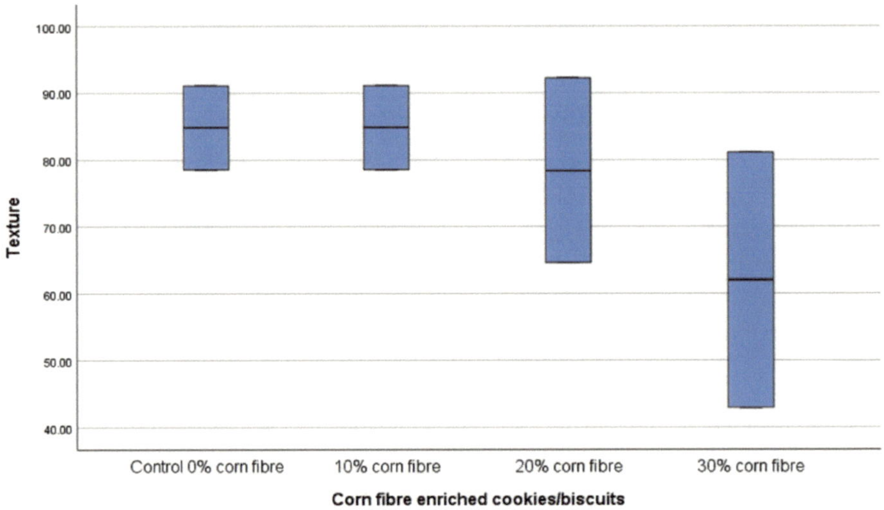

Figure 4.
Graph depicting the sensory attribute – the texture of corn fibre incorporated cookies/biscuits.

control sample, see **Table 7**. Incorporation of corn fibre at 30% gave cookies/biscuits the highest firmness 2.1 ± 0.2 (Kg Force).

In **Figure 4**, the texture of corn fibre enriched cookies/biscuits incrementally reduced with the incorporation of increased fibre up to 30%.

3.3.4 Rice bran enriched bread

In **Table 8**, incorporation of rice bran into up to 15% produced significantly ($p < 0.05$) undesirable overall acceptance. There was a significant difference ($p < 0.05$) in the aroma of the fibre enriched bread and control bread. All the rice bran fibre enriched bread was similar to each other but all significantly different ($p < 0.05$) from the control. There was no significant difference in the overall acceptance of 5% and 10% rice bran incorporated bread from the control sample.

In **Table 9**, there were no significant differences between rice bran supplemented bread up to 10%.

3.3.5 Sensory quality of corn bran enriched cakes

Based on sensory evaluation corn bran supplemented cakes up to 20% were found to be significantly desirable ($p < 0.05$) in comparison to control cake samples

Detail	Aroma	Crust colour	Crumb colour	Taste	Texture	Overall acceptance
Control (0%)	86.1 ± 3.9a	90.5 ± 2.4a	87.2 ± 3.9a	83.3 ± 1.6a	88.4 ± 0.8a	88.4 ± 0.8a
5% RB	73.9 ± 3.2ab	80.0 ± 7.7a	77.8 ± 6.4a	78.4 ± 3.2a	79.4 ± 5.8a	80.6 ± 0.6a
10% RB	71.1 ± 8.5ab	71.7 ± 11.6a	72.2 ± 8.5a	75.0 ± 6.7a	80.8 ± 13.6a	72.8 ± 10.4a
15% RB	68.9 ± 7.2ab	71.7 ± 11.3a	68.3 ± 6.5a*	68.9 ± 5.3a*	80.8 ± 8.3a*	68.4 ± 7.8a*

*Values are means ± standard deviation of sub-groups. Means in the same column with different superscripts are significantly different ($p < 0.05$). Key: RB – rice bran and * means 15% is significant in comparison to control only.*

Table 8.
Sensory quality of rice bran enriched bread.

Detail	Specific volume (ml/g)
Control (0%)	10.7 ± 9.1a
5% RB	10.1 ± 6.9a
10% RB	9.8 ± 6.6a

Values are means ± standard deviation of sub-groups. Means in the same column with different superscripts are significantly different ($p < 0.05$). Key: RB – rice bran.

Table 9.
Physical quality of rice bran enriched bread.

Detail	Crust colour	Taste	Texture	Overall acceptance
Control (0%)	87.6 ± 6.3a	81.2 ± 10.9a	85.3 ± 6.3a	83.2 ± 10.9a
10% CB	86.1 ± 2.3a	82.6 ± 10.4ab	85.3 ± 6.3a	84.4 ± 8.4ab
20% CB	80.6 ± 0.6b	79.3 ± 12.8ab	80.8 ± 13.6b	86.7 ± 5.3ab

Values are means ± standard deviation of sub-groups. Means in the same column with different superscripts are significantly different ($p < 0.05$). Key: CB – corn bran.

Table 10.
Sensory quality of corn bran enriched cakes.

for the attributes of crust colour, taste, texture and consequently obtained the overall acceptance, see **Table 10**.

3.3.6 Physical quality of corn bran enriched cakes

Corn bran incorporated cakes significantly ($p < 0.05$) impacted the crust luminosity of cakes at both 25% and 30% supplementation levels. All the other physical parameters were similar to the control sample. The texture, namely firmness increased as the level of corn bran increased in the cakes, see **Table 11**.

Initially, the texture of corn bran enriched cakes increased with the addition of corn bran, then reduced at 10% level of incorporation and thereafter incrementally increased as the supplementation levels were elevated in cakes. Not surprisingly, the highest degree of firmness in cakes is at the 30% level of corn bran incorporation in cakes, see **Figure 5**.

Detail	Crust L*	Crust a*	Crust b*	Crumb L*	Crumb a*	Crumb b*	Texture (Kg Force)
Control (0%)	84.4 ± 4.2a	−3.9 ± 6.9a	28.6 ± 11.6a	51.1 ± 17.9a	11.7 ± 3.7a	34.5 ± 14.2a	6.7 ± 1.5a
5% CB	80.8 ± 2.9a	−3.6 ± 6.6a	31.7 ± 6.4a	50.7 ± 15.3a	11.1 ± 3.2a	35.2 ± 11.7a	6.9 ± 1.2a
10% CB	78.1 ± 1.4a	−5.1 ± 6.2a	31.8 ± 7.6a	48.8 ± 17.5a	10.3 ± 2.6a	36.1 ± 9.8a	6.9 ± 0.9a
15% CB	73.9 ± 3.6a	−5.0 ± 6.6a	33.2 ± 6.5a	47.1 ± 16.8a	9.8 ± 4.1a	36.3 ± 9.2a	7.1 ± 1.1a
20% CB	71.9 ± 3.5a	−5.4 ± 7.5a	34.6 ± 5.6a	46.9 ± 17.7a	8.0 ± 3.8a	36.9 ± 7.4a	7.4 ± 1.2a
25% CB	68.2 ± 6.7b	−5.8 ± 8.5a	35.5 ± 5.2a	46.5 ± 18.4a	6.8 ± 4.6a	37.5 ± 6.6a	7.7 ± 1.2a
30% CB	64.7 ± 7.0b	36.6 ± 5.0a	36.6 ± 5.0a	45.7 ± 19.3a	5.4 ± 4.4a	37.5 ± 4.9a	7.7 ± 1.1a

Values are means ± standard deviation of sub-groups. Means in the same column with different superscripts are significantly different (p < 0.05). Key: CB – corn bran.

Table 11.
Physical quality of corn bran enriched cakes.

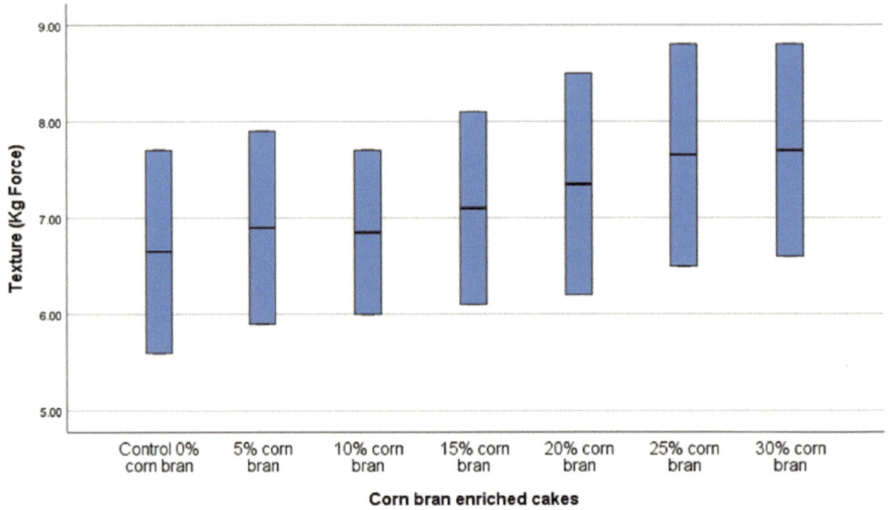

Figure 5.
Graph depicting the texture qualities of corn bran enriched cakes.

3.4 Data Characteristics and features

Several baked food products, namely cookies/biscuits, cakes and bread illustrated enhanced levels of dietary fibre, moisture, ash, minerals, vitamins contents and on the other hand reduction in proteins, fats and carbohydrates, see **Table 12**. The incorporation of rice bran up to the level of 15% into biscuits significantly ($p < 0.05$) enhanced its nutritional value [20]. In previous studies Yadav, Yadav, and Chaudhary [39] also found that the supplementation of biscuits up to 15% significantly ($p < 0.05$) enhanced its nutritional value in comparison to the control sample, namely significant increase in protein content from 7.3% to 15.4%; there was a non-significant increase in ash and fibre contents and a reduction in carbohydrates content. Several studies of cookies and biscuits using both higher rates of corn fibre (up to 40%) and rice bran (up to 20%) also reported a significant

Reference	Downs and Black quality assessment	Food vehicle	Fibre type and % incorporation	Control	n	Analysis parameters	Panel[a]	Outcomes
[45]	11/14	Cookies	Rice bran – acid stabilised and heat stabilised (0–20%)	Wheat flour	5	Physical parameters: • Width (mm) • Thickness (mm) • Spread factor (%) • Texture (Kg F) Sensory attributes: flavour, colour, appearance, taste, texture and overall acceptability.	Postgraduate students – Food Technology Department (T)	• No significant difference was detected in the chemical and physical properties of cookies incorporated with acid stabilised rice bran (ASRB) and heat stabilised rice bran (HSRB). • The incorporation of brans in cookies resulted in a significant incremental increase of moisture, crude protein, fat and mineral contents. • The parameters of average width, thickness and spread factor of cookies were also increased as rice brans were added. • The scores for colour of cookies decreased significantly as the level of rice bran increased, but not was non-significant at the 10 percent level of substitution.
[20]	11/14	Biscuits	Rice bran (0–10%)	Wheat flour	26	Physical parameters: • Width (mm) • Thickness (mm) • Spread factor (%) • Texture (Kg F) Sensory attributes: flavour, colour, appearance, taste, texture and overall acceptability.	Males and females (50% males and 50% females), age ranged 20–50 years, untrained panellists (UT)	• Biscuits supplemented with 5–10% heat stabilised rice bran using flour containing (rice flour 35% + maize flour 35% + pea flour 30%) resulted in desirable overall acceptance by the sensory panel. • Incremental addition of stabilised rice bran significantly increased moisture, ash, crude protein, fibre and thickness of the biscuits. • Width and spread factor of biscuits reduced with increasing levels of stabilised rice bran.
[39]	12/14	Biscuit	Rice bran protein concentrate	Refined wheat flour	10	Physical parameters: • Width (mm) • Thickness (mm) • Spread factor (%)	Semi-trained members (ST)	• The substitution of refined wheat flours up to 10 per cent using RBPC resulted in the formulation of protein-enriched biscuits with favourable overall acceptability.

Reference	Downs and Black quality assessment	Food vehicle	Fibre type and % incorporation	Control	n	Analysis parameters	Panel[a]	Outcomes
			(RBPC) (0–15%)			• Texture (Kg F) Sensory attributes: flavour, colour, appearance, taste, texture and overall acceptability.		• There was a significant increase in the protein content of biscuits produced from 7.3% in control biscuits to 15.4% in the biscuits with 15% rice bran supplementation. • The fracture strength was also significantly higher than that of the control biscuits ($p < 0.05$).
[46]	12/14	Biscuits	Corn fibre (0–30%)	Wheat flour	60	• Texture • Diameter (W):mm • Thickness (T): mm • Spread ratio • % spread factor • Sensory attributes of appearance, colour, texture, taste, flavour and overall acceptability.	Untrained panellists	• Significant enhancement of protein and dietary fibre contents of the cookies. • Hardness of the cookies also increased as the level of dried young corn increased in the cookies. • Supplementation of cookies with 10% dietary obtained the best scores in comparison to the control.
[36]	10/14	Biscuits	Corn fibre (10–40%)	Refined wheat flour	10	• Texture • Diameter (W):mm • Thickness (T): mm • Spread ratio • % spread factor • Sensory attributes of appearance, colour, texture, taste, flavour and overall acceptability.	Semi-trained panellists (ST)	• Corn fibre replacement in refined wheat flour resulted in enhanced hardness, thickness, moisture, total dietary fibre (TDF) and moisture in biscuits and simultaneously a significant reduction in carbohydrates and energy. • Reduced diameter was observed and a reduced spread ratio was derived. • Sensory evaluation results of the attributes of colour, appearance, texture, taste, flavour and overall acceptability of biscuits were similar to the control up to the level of 20% addition of corn fibre.

Reference	Downs and Black quality assessment	Food vehicle	Fibre type and % incorporation	Control	n	Analysis parameters	Panel[a]	Outcomes
[38]	11/14	Cookies	Sugarcane bagasse – with peel (0–10%)	Wheat flour	30	Physical parameters: • Width (mm) • Thickness (mm) • Spread factor (%) • Colour: L*, a* and b* • Texture (Kg F) Sensory attributes: colour, appearance, taste, texture and overall acceptability.	Trained panel members (T)	• The incorporation of sugarcane bagasse with a peel at 5% to enrich cookies displayed the highest overall acceptability. • Enriched cookies with 5% peeled sugarcane contained increased phenolic content, moisture content, fat and ash content, but slightly lower protein content in comparison to the control. • 5% enriched sugarcane bagasse (with peel) possessed a golden-brown colour, along with increased hardness, diameter and spread ratio but reduced thickness in comparison to the control cookies. Both cookies were of similar weight.
[37]	9/14	Biscuits	Sugarcane bagasse (0–15%)	Wheat flour	6	Physical parameters: • Width (mm) • Thickness (mm) • Spread factor (%) • Colour: L*, a* and b* • Texture (Kg F) Sensory attributes: colour, appearance, taste, texture and overall acceptability.	Panel members (UT)	• Biscuits with the incorporation of 10% steamed sugarcane bagasse and additives were highly acceptable based on the results of sensory evaluation. • Hardness and moisture of fibre enriched biscuits increased as the percentage of fibre added increased, contrastingly the surface features decreased. At 10% level of addition with additives, the texture of biscuits was enhanced and surface features improved. • The colour of the crumb at 10% fibre addition with additives was similar to the control – pale to yellow-brown, also its spread value and width increased, but thickness decreased.

Reference	Downs and Black quality assessment	Food vehicle	Fibre type and % incorporation	Control	n	Analysis parameters	Panel[a]	Outcomes
								• Dietary fibre content increased in fibre enriched biscuits but protein and fat contents were reduced in comparison to the control. • Micro-bacterial load of bagasse was reduced through the steaming process, and rheological features such as dough development, stability and viscosity were all negatively affected with increasing levels of fibre addition above 10%.
[47]	11/14	Bread	Full fatted rice bran (0–15%)	Wheat flour	20	Physical parameters: • Volume (ml) • Mass (g) • Specific volume (g/ml) Sensory attributes: aroma, crust colour, crumb colour, taste firmness, taste and overall acceptability.	Staff members and students (T)	• There were significant percentage increases in the nutrient contents of the rice bran supplemented bread, most notably the increase in protein, crude fat, carbohydrates, moisture, minerals and vitamins. • Importantly, there was a reduction in the sodium content of the bread samples. • As the level of rice bran incorporation increased in the bread samples from 5–15% it resulted in enhanced bread weight. • Contrastingly, both the volume and specific volumes of the bread samples were reduced. • There were significant differences illustrated between the physical properties of the control bread samples and the fibre incorporated bread samples: in both cases of dough and bread loaves.

Reference	Downs and Black quality assessment	Food vehicle	Fibre type and % incorporation	Control	n	Analysis parameters	Panel[a]	Outcomes
[31]	11/14	Bread	Full fatted & defatted rice bran (0–15%)	Wheat flour	10	Physical parameters: • Volume (ml) • Mass (g) • Specific volume (g/ml) Sensory attributes: aroma, crust colour, crumb colour, taste firmness, taste and overall acceptability.	Staff members and students (T)	• The hardness of composite bread increased with elevated supplementation of rice bran in flour. • The specific volume of the composite bread reduced with the addition of rice bran from 5–15%. • Rice bran incorporation in wheat flour enhanced the protein and crude fibre in comparison to the control wheat flour bread. • Supplementation of wheat flour with 10% of full fatted and defatted rice bran in the preparation of composite bread was found to be desirable overall • acceptability when compared to the control composite bread.
[48]	12/14	Bread	Maize bran (0%&30%)	Wheat flour	30	Physical parameters: • Volume (ml) • Mass (g) • Specific volume (g/ml) • Colour: L^*, a^* and b^* Sensory attributes: flavour, crust colour, crumb colour, colour, taste, texture, and overall acceptability.	Untrained tasters (UT)	• The incorporation of brans in bread formula resulted in reduced bread volumes, and enhanced water, ash, lipids, proteins, fibres and phytates contents. • Maize bran supplemented bread recorded the highest phytate value (0.53 g/100 g DM) although lower than the control (2.5 g/100 g DM). • The addition of maize significantly ($p < 0.05$) enhanced mineral contents such as magnesium, potassium, zinc, manganese and iron in comparison to the control bread but had no significant effect on calcium and copper contents.

Reference	Downs and Black quality assessment	Food vehicle	Fibre type and % incorporation	Control	n	Analysis parameters	Panel[a]	Outcomes
[44]	8/14	Bread (chapatti/ fermented brad)	Sugarcane bagasse (0, 5 & 8%)	Wheat flour	50	Physical parameters: ND Sensory attributes: flavour, aroma, crust colour, and texture. Overall acceptability: ND	Students, between the ages of 18 and 36 years old (ST)	• Hedonic responses suggested that the addition of 8 wt.% of treated fibres during the production of chapatti-type fermented bread (with SCB) is suitable for human consumption. • Incorporation of food with sugarcane bagasse enhanced its nutritional value. • Increased in total fibre 7.4 ± 0.5–11.7 ± 0.6 g/100 g. • Similarity in the inhibition activity of the 2–2 diphenyl-1 picrylhydrazyl (DPPH) free radical values between control and sugarcane bagasse enriched bread. • Lower levels of crude fat, protein and ash content in sugarcane bagasse enriched chapatti bread in comparison to control samples and the other sugarcane tops enriched samples.
[17]	14/14	Cake	Corn bran (0–30%)	Cake formulation containing 0% corn bran	30	Physical parameters: • Crumb colour: L^*, a^* and b^* • Crust colour: • L^*, a^* and b^* • Texture (Kg F) Sensory attributes: flavour, crust colour, crumb colour, taste and overall acceptability.	Trained taste panellists (T)	• The inclusion of 5–10% modified corn bran in the cake formulation gave rise to the production of a functional cake. Above 10% fibre addition resulted in increased firmness and gumminess, but reduced springiness and cohesiveness. The colour of cakes significantly ($p < 0.5$) increased as corn bran incorporation levels rise above 5%. • Best acceptability of corn bran was at 5% followed by 10% supplementation levels. • 30% incorporation of corn bran obtained the lowest acceptability.

Reference	Downs and Black quality assessment	Food vehicle	Fibre type and % incorporation	Control	n	Analysis parameters	Panel[a]	Outcomes
[49]	13/14	Banana cakes	Sugarcane bagasse (0–6%)	Oat flour	50	Physical parameters: • Crumb colour: L*, a* and b* • Crust colour: • L*, a* and b* • Texture (Kg F) Sensory attributes: crust colour, taste and overall acceptability.	Untrained panel members: mixture of staff and students (UT)	• The addition of sugarcane bagasse resulted in no distinct alteration in the chemical composition of the cakes. • Increased dietary fibre incorporation levels in cake resulted in increased firmness of the cake. • There was no interference in the sensorial acceptance of the product when dietary fibre was added between 3 and 6%. • Cakes with 6% sugarcane bagasse incorporation were considered acceptable and resulted in the preparation of desirable high fibre cakes.
[23]	12/14	Cakes	Corn bran (0–30%)	Cake flour	25	Physical parameters: • Crumb colour: L*, a* and b* • Crust colour: • L*, a* and b* • Texture (Kg F) Sensory attributes: flavour, crust colour, crumb colour, taste and overall acceptability.	Untrained panellists (UT)	• The increasing levels of corn bran replacement in cake batter showed no influence on the hardness and springiness of cakes. • Substitution of flour with 20% corn bran resulted in cakes with favourable sensory scores in the attributes of texture, taste and overall acceptability of the cakes.

[a]*Description is given by author. Trained (T), semi-trained (ST) and untrained (UT).*

Table 12.
General study features and characteristics of cookies/biscuits, bread and cakes.

increase in protein, fibre, minerals and moisture contents, on the contrary, fat contents and total carbohydrates showed a reduction in cookies and biscuits [45, 46].

In another study, protein contents in corn fibre enriched biscuits showed a reduction, as reported in most other studies ash, moisture and total dietary fibre contents were higher than in control biscuits [36]. Jauharah, Wan Ishak, and Robert [46] indicated that supplementation of biscuits with corn fibre up to 30% resulted in enhanced energy value. Contrastingly, more recently Sandhu, Bains, Singla, and Sangwan [36] revealed that biscuits supplemented with corn fibre up to 40% produced enriched biscuits with reduced energy value in comparison to control (a reduction from 499 Kcal to 486 Kcal). Moreover, there was evidence of reduced carbohydrates contents in corn fibre incorporated biscuits [36]. A recent study of sugarcane bagasse incorporation of cookies up to 10% Vijerathna et al. [38] indicated the presence of enhanced phenolic content and significantly higher ash and moisture contents, marginally similar fat composition in control and enriched cookies, but protein contents in enriched cookies were significantly lower than in the control sample. In 2011, Sangeetha, Mahadevamma, Begum, and Sudha indicated that sugarcane bagasse supplemented biscuits up to 15% biscuits possessed reduced total fat and protein contents, meanwhile, there were minimal variances in moisture content, ash and acid-insoluble contents in comparison to control.

There were two (2) studies, namely [17, 23] relating to corn bran supplemented cakes using similar incorporation percentages, namely 0–30% and that did not conduct any nutritional analysis, hence no data nutritional data were extracted from these studies. Rice bran supplemented bread up to 15% resulted in a significant (0.05) increase in moisture, protein, crude fibre, crude fat, ash and several minerals and vitamins in comparison to control bread [31, 47]. Importantly, it was reported in a study of enriched bread with rice bran that the sodium content was significantly (0.05) reduced, the composition of carbohydrates was reduced in enriched bread [47]. Further, a significant increase ($p < 0.05$) was reported in ash, moisture, proteins, lipids and minerals for example magnesium, potassium, zinc, manganese and iron; carbohydrates and energy value (Kcal/100 g) was significantly reduced ($p < 0.05$) and interestingly phytic acid contents of enriched bread using maize bran at 30% supplementation [48]. More recently, Gil-López et al. [44] stated that supplementation of chapatti using sugarcane bagasse resulted in enhanced total fibre 7.4 ± 0.5–11.7 ± 0.6 g/100 g, promote inhibition activity of the 2–2 diphenyl-1 picrylhydrazyl (DPPH) and also reduced the levels of crude fat, protein and ash content in enriched bread in comparison to control samples.

4. Discussion

4.1 Roles of dietary fibre in foods

4.1.1 Fortification of foods

Dietary arabinoxylan-based sources namely, rice bran, corn, and sugarcane bagasse can be used to improve the nutritional contents of baked foods and making them become functional products. In recent studies, Haghighi-Manesh & Azizi [17] reported that the inclusion of 5–10% modified corn bran in the cake formulation gave rise to the production of a functional cake with reduced cohesiveness and springiness higher gumminess, darkness, and favourable sensory properties. This suggests that corn bran can be added to baked food products to enhance nutritive value, textural qualities and sensory qualities. Moreover, incorporation of food with sugarcane bagasse at 8% supplementation and other dietary fibre arabinoxylan sources into food systems such as bread tend to enhance its nutritional value and influence changes in the physical, rheological, and sensorial characteristics of foods [24, 44]. Also, previously, Amna, Bhatti, Anwaar, & Randhawa [45] indicated that enhance moisture, protein, fat and minerals (calcium, manganese, and magnesium) composition in cookies was reported as the level of rice bran incorporation increase in biscuits. The evidence suggests that 20% fibre incorporated rice bran cookies may possess a significant increase in moisture, protein and minerals (zinc and Iron) contents. Meanwhile, in bread, Pauline et al., [48] indicated a significant increase ($p < 0.05$) in water, ash, lipids, proteins, fibres and phytic acid contents between the control sample and enriched maize bran bread. The incorporation of 8.2 g of AXE per 100 g of available carbohydrates into bread facilitated the criteria for the health claim of the reduction of post-prandial glycaemic response [7, 50]. This suggests that enrichment using arabinoxylans-based sources can improve the nutritive value of ordinary food products.

4.1.2 Water retention capacity

Arabinoxylans dietary fibre sources tend to affect the moisture content of foods positively. Several studies reported the increase of moisture content in the enrichment of cookies/biscuits, bread and cakes, in comparison to the control samples, in some cases moisture was increased significantly ($p < 0.005$) and other cases marginally [31, 36, 38, 48, 49]. Javaria [20] indicated that moisture content in cookies/biscuits can consider a fundamental quality since it tends to influence both end quality and shelf life of these food products. Arabinoxylan based sources during addition to dough tend to enhance viscosity and water absorption; improved dough development, reduce starch retrogradation and reduce the firmness of foods due to the presence of bound water facilitating less stiff gluten and starch network [22, 51]. This suggests that arabinoxylan dietary fibres supplementation into foods can enhance its moisture content and affect other properties in foods namely its storage properties and shelf life. However, in bread, it was proven that arabinoxylan has a beneficial effect on water activity and thus favourably influenced bread freshness [22]. Moreover, Jauharah, Wan Ishak, & Robert [46] stated that moisture in the range of 1–5% is considered a benchmark for cookies/biscuits and that fresh corn fibre may lead to too high moisture levels in foods and consequently result in food spoilage. It can be suggested that excess moisture may result in micro-bacterial activities which can impact shelf life unfavourably. Similarly, in the enrichment of bread with 20% corn bran high, bread quality was significantly affected by the water content in the composite formula [52]. Taken together, water retention capacity is fundamental in influencing baked foods' end quality.

4.1.3 Dietary fibre enriched food qualities and functional food ingredients

Arabinoxylans sources, namely rice bran, corn fibre and sugarcane bagasse tend to influence various qualities of enriched foods. Arabinoxylans (AXE) contribute essentially in determining the physical and chemical properties of the final quality of food products and made up a high percentage of the cell walls of cereal grains and also present in sugarcane bagasse [16, 22, 53]. In earlier studies, Foschia, Peressini, Sensidoni, and Brennan [13] indicated also that the quality and nutritional aspects of cereal products are influenced by the addition of dietary fibres into foods. Importantly, rheological properties among arabinoxylan fibres such as corn fibres displayed differences that may influence the quality of food products after their incorporation during food processing [54]. It can be hypothesised that rheological factors can contribute to some of the undesirable features in dietary fibre enrichment of baked foods. Numerous studies have conducted dietary fibre supplementation of cookies/biscuits, bread and cakes using rice bran, corn fibre and sugarcane bagasse at different rates of incorporation and with varying results, see **Table 12** [17, 23, 38, 39, 47]. This suggests that fibre incorporation using rice bran, corn fibre and sugarcane bagasse tend to produce both desirable and undesirable physical and sensorial features in baked food products.

Moreover, because of the favourable nutritive value impact of dietary fibres on enriched foods, and the health-related benefits that be obtained upon consumption in a prescribed manner, it can be suggested that these ingredients can potentially be considered as functional ingredients in baked foods. In support of the suggestion, Zidan and Eldemery [55] found that the incorporation of 5–10% defatted black rice bran was found to possess acceptable sensory features and resulting in the nutritive value of bread being enhanced with minerals such as phosphorus, potassium, iron, copper, zinc and calcium. Contrastingly, Zhang et al. [40] found that arabinoxylan fortification of bread with 10% arabinoxylan fibre illustrated undesirable physical qualities in bread such as decreased specific volume, harder crumb, darker crust colour, and a coarser crumb structure. Overall dietary fibres facilitate the production of enriched foods, which may possess both desirable and undesirable sensorial and physical qualities, which can be attributed to the rheological behaviour of dietary fibre arabinoxylan when used in food supplementation.

4.2 Effects of the incorporation of dietary fibre derived from selected cereals (rice bran and corn fibre) and sugarcane bagasse on the physical and sensory qualities of baked food products: cookies/biscuits, bread and cakes

The incorporation of sugarcane bagasse at both 5% and 10% in cookies/biscuits resulted in significantly undesirable ($p < 0.05$) overall acceptance in sensory qualities, namely appearance, taste and texture, (see **Table 2**). It can be suggested that the increased firmness of the cookies/biscuits increased as the rate of supplementation of sugarcane bagasse increased affected its overall acceptance negatively, (see **Table 3**). Javaria [20] indicated that the firmness in cookies tends to influence their overall acceptance by the panellists. Sugarcane bagasse incorporation in cookies / biscuits showed no significant improvement in the physical qualities of the enriched product in comparison to the control. This can be attributed to the rheological properties of dietary arabinoxylans within the food matrix of the enriched biscuits. Kale, Pai, Hamaker, and Campanella [54] found that the rheological properties of extensional viscosity and solution viscosity of arabinoxylans fibres in the food system tend to influence its quality. This suggests that as the competition may have increased between the dietary fibre and the protein network, it resulted in a re-arrangement of the food matrix network. The incremental addition of sugarcane bagasse resulted in at least 50% rise in the firmness of 10% enriched cookies/biscuits. This can be attributed to the stiffness of the gluten and starch matrix may have been destroyed due to the competition of unextracted arabinoxylan dietary fibre and proteins for bound water. In a previous study of biscuits, Sozer, Cicerelli, Heiniö, and Poutanen [56] found that the hardness of biscuits was increased due to the particle size reduction of bran, while the starch hydrolysis index of biscuits decreased due to particle size reduction of bran.

In rice bran enriched cookies/biscuits incorporation levels of 5–15% showed a significant difference in comparison to control cookies/biscuits. Interestingly, a low fibre incorporation of rice bran of 5% displayed significantly ($p < 0.05$) undesirable acceptance, similar to 10% and 15% rice bran enriched cookies/biscuits, (see **Table 4**). These results agree with previous studies that reported a reduction in overall acceptance scores in enriched foods as the fibre incorporation increased [20, 45]. Contrastingly, the addition of 5% dietary fibre was acceptable based on sensory evaluation results in wheat bran enriched biscuits. Evidence suggests that rice bran significantly ($p < 0.05$) enhanced the thickness of cookies/biscuits, however, other physical qualities such as width and spread factor (SF%) showed no significant differences ($p > 0.05$), (see **Table 5**). Similar, findings were reported in some other fibre supplementation studies conducted. This can be as a result of the enhanced moisture absorption capacity of the dietary fibre used in the enriched cookies. In a recent study, Vijerathna et al. [38] reported that the enhanced size of the hydrophilic starch granules increased the thickness of fibre enriched cookies.

Corn fibre supplemented cookies displayed desirable overall acceptance at 20% incorporation rate and was significantly different ($p < 0.05$) in comparison to control cookies/biscuits, as a result, enriched corn fibre cookies/biscuit obtained the highest sensory score, (see **Table 6**). This suggests that corn fibre can be incorporated at a higher rate and produced desirable results possibly due to its high water retention capacity and its particle. In previous studies, Mishra & Chandra [57] reported a similar trend in results using a higher incorporation rate in cookies/biscuits. Similarly, supplementation of cookies/biscuits at 10% and 20% were significantly different ($p < 0.05$) to control, 20% corn fibre cookies/biscuits followed by 10% supplemented corn fibre cookies/biscuits obtained the best scores during panel sensory evaluation. This desirable overall acceptance at 10% - 20% incorporation may have been influenced by cookies/biscuits increase moistures that facilitated texture 1.8 ± 0.5 (Kg Force) and 1.9 ± 0.4 (Kg Force) at 10% and 20% respectively; and colour of the enriched product. Cookies/biscuits incorporated with corn fibre at 30% obtained the lowest overall acceptance score. This suggests that increased firmness and other sensory attributes such as flavour, colour, appearance and texture acceptance which were significantly undesirable ($p < 0.05$) based on sensory panel evaluation contributed to a low overall acceptance of the food product.

In terms of physical parameters, results suggest that the firmness of corn fibre enriched cookies/biscuits increased with increasing incorporation of corn fibre up to 30%. Jia et al. [58] indicated that water-insoluble proteins in flours that possessed minimal gluten, namely glutenin and gliadin combined to influence elasticity and structural strength of dough (glutenin) and viscosity and fluidity of dough (gliadin) tend to intertwine and formed a strong protein network structure in the food matrix. This suggests that the degree of firmness affected the protein network in the food matrix negatively possibly due to low moisture levels which contributed greatly to the high texture of the cookies/biscuits [52]. Cookies/biscuits (51.9 ± 5.0 mm) supplemented at 30% were thicker than the control (49.1 ± 24.5), however, there was no significant differences ($p > 0.05$), (see **Table 7**). This can be attributed to the high moisture absorption capacity of corn fibre, a satisfactory source of arabinoxylan dietary fibre which possibly clings to the bound water within the food matrix network, as the viscosity of the solution rises in the food system, leaving the protein components to form strong aggregates. The polysaccharide, arabinoxylan tends to possess various physicochemical properties, such as the high capacity to retain water and display an inclination to form high viscosity solutions when incorporated into the food matrix [22].

Incorporation of rice bran into bread at 15% showed a significant undesirable overall acceptance ($p < 0.05$) in comparison to control samples for all sensory attributes, except crust colour, (see **Table 8**). This can be attributed to a possible darker crumb in comparison to the control. Ortiz de Erive, Wang, He, and Chen [52] indicated that due to the enrichment of bread with dietary fibre their colour is likely to be darker in comparison to the non-enriched sample, as a result of the lower baking temperature of the crumb, thus affecting caramelization or Maillard's reaction. Arabinoxylans significantly affected protein network formation influencing the food matrix features and aroma of incorporated food products [59]. The aroma of enriched bread up to 10% was significantly different ($p < 0.05$) from the control. Enriched rice bran at 15% obtained the lowest overall acceptance score by the sensory evaluation panellists. In other studies, rice bran enriched bread was found to be acceptable at 5–10% [55]. This may be due to the sensory panel members' dietary familiarity with wheat flour bread. There was no significant difference ($p > 0.05$) in the specific volume (ml/g) between enriched rice bran bread and the control, however, specific volume reduced as rice bran incorporation percentage increase in the enriched bread sample, (see **Table 9**). This can be attributed to the water absorption capacity of the rice bran, which absorbs potentially most of the moisture, its ability to form viscous solutions, thus leaving the gluten component in the bread matrix inadequately hydrated. Pauline et al. [48] reported that rice bran enriched showed a lower specific volume in comparison to the enriched bread. Moreover, as rice bran is being incorporated into the enriched bread, due to the high-water capacity of rice bran, the enriched bread texture is likely to increase and become denser and less porous, thus resulting in the continual reduction of the specific volume from 10.7 ± 9.1 at 0% rice bran incorporation to 9.8 ± 6.6 at 10% rice bran incorporation [52]. It can be suggested that textural qualities of bread tend tends to be affected by its moisture content.

Corn bran enriched cakes up to 20% fibre incorporation displayed significantly ($p < 0.05$) more desirable sensory qualities than control cakes, (see **Table 10**). This can be attributed to the increased moisture content levels caused by the addition of corn bran during the enrichment process, which allowed the corn bran supplemented cakes to be adequately hydrated. There was a significant difference in crust colour of corn bran enriched cakes and control at the 20% fibre incorporation level. This can be attributed to the possible pale colour of the enriched cakes at the 20% level of fibre incorporation. Generally, dietary fibre enriched cakes take the inherent colour of the fibre being utilised in the process of enrichment, until Maillard's reaction occurs [52]. The increasing levels of corn bran replacement in cake batter showed no influence on the hardness and springiness of cakes [23]. In terms of physical qualities, the firmness of cakes increased incrementally with the increase in corn bran supplementation, (see **Table 11**). Moreover, there was no significant difference ($p < 0.05$) in physical qualities, except in crust L^* at 25% and 30% enrichment. This can be attributed to the possible darker colour of the corn bran enriched cakes crust in comparison to the control. The high baking temperature will tend to affect the fibre enriched cakes' crust due to caramelization or Maillard's reaction and the inherent colour of the fibre [52]. In summary, the incorporation of dietary fibres, rice bran and sugarcane bagasse affected the sensory and physical qualities of baked food products unfavourably. However, corn fibre and corn bran produced desirable sensory and physical effects at 20% fibre supplementation in both cookies/biscuits and cakes.

4.3 Beneficial implications of dietary fibres in foods

4.3.1 Nutritive value of enriched foods

The fundamental evidence of this review suggests that as the level of fibre incorporation increased the nutritive value of the food products cookies/biscuits, bread and cakes were significantly ($p < 0.05$) enhanced, resulting in increased moisture, ash, protein, minerals, vitamins, dietary fibres and crude fats, essentially in cereal-based incorporated food products. Corn bran enriched bread was suggested to contain significant ($p < 0.05$) contents of phytic acid contents. More recently, Ekpa, Palacios-Rojas, Kruseman, Fogliano, & Linnemann [4] indicated that food enrichment is an essential processing technique to enhance the nutrient content of staple foods including baked food products, thus, this suggests that fibre enriched foods can potentially provide convenience to consumers and ultimately facilitates improved food nutrition guidelines for the society and foster food security locally. Similarly, rice bran possessed vital antioxidants, which comprised of the well-recognised immune system enhancing compound, namely phytosterols; polysaccharides; minerals and trace minerals including magnesium, selenium, zinc, vitamin E, omega-3 fatty acids and many other phytonutrients [28]. This suggests that the strategy of fibre enrichment of popular food products can potentially be utilised more widely by stakeholders in the food industry in collaboration with health care agencies of various age groups, stakeholders of feeding programmes in schools and policymakers in governmental public health agencies. In another recent study, corn fibre was found to be an effective antioxidant than wheat bran and this was attributed to its elevated ferulic acid content and polyamine-conjugates. Interestingly, hydroxycinnamates may not be necessary for the antioxidant effect [30]. Contrastingly, in sugarcane bagasse supplemented food products, there were instances of reduced total fat, protein and marginally increased in moisture content of food products. Reduction in protein and fats contents was also reported in other fibre supplementation research [60]. Overall, the nutritive value of food products can be enhanced by fibre enrichment.

4.3.2 Health benefits

It has been widely established that numerous health benefits are associated with the consumption of enriched food products in recommended administered quantities. Moreover, dietary fibre potentially can contribute to several health benefits such as enhanced bowel function, reduced levels of cholesterol in the body, better weight maintenance and assisted in controlling blood sugar levels in the human body [24]. Chen et al. [5] indicated that there are several beneficial impacts on various physiological processes which can be attributed to arabinoxylans and consequently health functions and prebiotic effects are being influenced by arabinoxylans structures, which in turn depends on its method of extraction and source. This suggests that rice bran, corn fibre and sugarcane bagasse, which are well-established sources of dietary fibre arabinoxylans, can potentially be utilised to produce a wider range of enriched foods which can confer various types of health benefits to the consumers. However, the consumption of dietary fibre should adhere to the established guideline of at least 25 g/day for adults [6].

Dietary fibre supplemented foods with rice bran, corn fibre and sugarcane bagasse, namely cookies/biscuits, bread and cakes enriched foods was significantly ($p < 0.05$) higher in nutritive value, and contrastingly tend to contain significantly ($p < 0.05$) reduced energy value (Kcal/100 g) in comparison to control samples. Evidence suggests that fibre enriched cookies/biscuits, bread and cakes were significantly enhanced ($p < 0.05$) by supplementation, thus they can be considered potentially as functional foods. Moreover, further evidence indicated that supplementation of chapatti using sugarcane bagasse resulted in enhanced total fibre 7.4 ± 0.5–11.7 ± 0.6 g/100 g and contributed to elevated inhibition activity of the 2–2 diphenyl-1 picrylhydrazyl (DPPH) free radical values. Diets with enhanced fibre contents are associated with desirable effects on the health of consumers [3]. In general, fibre enriched food products have shown a reduction in energy levels in comparison to the control samples. This suggests that enriched foods can be used for individuals with varying health conditions and risk factors such as coeliac patients and other different target groups.

4.3.3 Food policy

Dietary fibre enriched foods can potentially be utilised in various governmental-based food policies. Recently, Lockyer and Spiro [61] reported that the average fibre intake in the United Kingdom can be considered fairly below the level recommended level. This suggests that a fibre enrichment strategy involving the relevant stakeholders can potentially facilitate an improvement in dietary fibre intake within the population. Previously, a comprehensive review study was conducted for modern dietary and policy priorities for cardiovascular diseases, obesity, and diabetes and found that there are complex influences of different foods on long-term weight regulation and recommended implementing an evidence-based strategy, including policy approaches, for lifestyle changes [62]. Importantly, fibre enriched foods contain rice bran, corn fibre or sugarcane bagasse all can be used in producing functional foods suitable for the particular health setting, thus this indicates enriched foods can be utilised in different food policies with different outcomes.

4.3.4 Consumer acceptance

Despite, the promising nutritional enhancement of enriched food products, rice bran and sugarcane bagasse at increasing incorporation levels tend to be associated with undesirable sensory and physical qualities of baked foods [55]. Evidence of this review suggests that lower incorporation rates can potentially result in improved overall acceptance of fibre enriched food products, further results suggest that the level of overall acceptance of baked products cookies/biscuits, bread and cakes were considered unacceptable above 20% incorporation for rice bran and sugarcane bagasse any type particular baked product. In support of evidence, [39] stated that incorporation of wheat flour using 10% rice bran protein concentrate (RBPC) resulted in the production of protein-enriched biscuits with favourable overall acceptability. It can be suggested that consumers tend to have a potentially better awareness of the health benefits of dietary fibre enriched foods, and also some of the most popular products are being enriched with dietary fibre thus creating a potential health trend among consumers. Similarly, Gul, Yousuf, Singh, Singh, & Wani [32] stated that consumer's attitude towards healthy foods is improving and thus presents potential opportunities for further development of functional foods on the world markets. This suggests dietary fibre food ingredients can potentially find wide applications in the other fields, thus increasing consumer awareness of dietary fibre enriched foods. In summation, using popular foods such as enriched bread and other potentially widely consumed food products as the benchmark for the enrichment of other products can facilitate further acceptance of consumers.

4.4 Limitations of study

There were three (3) main limitations that were experienced in the conduct of this review, namely the framework of the search strategy which influenced the availability of an adequate number and relevant primary research articles relating to the incorporation of sugarcane bagasse in baked food products such as cookies/biscuits, bread and cakes; the appropriateness of primary research articles to meet the selection criteria for inclusion in the review for analysis and the high percentage of primary research articles relating to dietary fibre supplementation in baked foods originating from predominantly less developed and developing countries based in Asia such as India, Pakistan, Bangladesh, Malaysia, Iran and Sri Lanka. The framework of the search strategy returned a relatively high percentage of primary research relating to wheat bran incorporation in foods, other cereal fibres such as barley, rye and oats, fibres from fruits and vegetable parts among others. Approximately less than 5% of the primary research articles were related to the incorporation of sugarcane bagasse in baked foods.

The appropriateness of primary research articles to meet the selection criteria for inclusion in the review for analysis resulted in some potentially interesting primary research studies not being include in the review, namely [7, 35, 59, 63]. Moreover, some of these studies was unable to be placed into sub-groups to facilitate statistical analysis such as [44, 48, 49] among others and therefore there were excluded from the list of included studies of the review. Moreover, primary research studies which utilised arabinoxylan extract as the source of incorporation in baked foods were extremely minimal, namely [35, 63], however both failed to achieved the selection criteria and thus were not selected. Therefore, the analysis of studies with arabinoxylan extract supplemented at a lower percentage was not possible.

The majority of primary research articles originated from developing and less developing countries located in Asia such as [20, 37–39, 45] among others. Thus, studies were able to be analysed from a wider cross-section of laboratory settings.

5. Conclusion

This systematic review demonstrates the utilisation of a comprehensive research methodology in the selection and examination of fourteen (14) dietary fibre food supplementation primary research studies to provide relevant and impartial new insights of the effects of the incorporation of dietary fibre derived from selected cereals (rice bran and corn fibre) and sugarcane bagasse on the physical and sensory qualities of baked food products: cookies/biscuits, bread and cakes. Arabinoxylans-based dietary fibre sources' roles in food supplementation involves enhancing the nutritive value of ordinary food products; influencing the end quality of baked foods and potentially improving the sensory and physical qualities of baked foods. The supplementation of sugarcane bagasse at both 5% and 10% and rice bran up to 15% into cookies/biscuits resulted in significantly undesirable acceptance ($p < 0.05$). Corn fibre was supplemented into cookies/biscuits up to 20% and had a favourably significant ($p < 0.05$) impact on its sensory qualities. It was suggested that enhance moisture content from corn fibre incorporation in combination with its particle size contributed to this desirable outcome.

Sugarcane bagasse incorporation negatively affected the physical qualities of cookies/biscuits. The incremental addition of sugarcane bagasse resulted in a 50% rise in the firmness of 10% enriched cookies/biscuits, from 5.7 ± 5.4 (Kg Force) to 13.0 ± 3.9 (Kg Force). Rice bran significantly increase ($p < 0.05$) the thickness of cookies/biscuits, from 8.6 ± 1.0 (mm) to 9.9 ± 0.3 (mm), the width and spread factor were similar to control. Corn fibre cookies supplementation did not significantly affect its physical qualities. Rice bran incorporation into bread at 15% showed a significant ($p < 0.05$) undesirable effect on its sensory qualities. However, the was no significant adverse effect on its physical quality. Corn bran enriched cakes up to 20% fibre incorporation displayed a significant ($p < 0.05$) favourable effect on the sensory properties of cakes; contrastingly, it resulted in a significant undesirable physical effect on the crust colour of corn bran enriched cakes.

There were four (4) main beneficial implications of dietary fibre food fortification using rice bran, corn fibre and sugarcane bagasse, namely the evidence suggests the enhancement of the nutritive value of foods; allows for the potential production of fibre enrichment foods to cater for particular target groups; dietary fibre enriched foods can be utilised in various food policies with particular outcomes, such as increasing the fibre intake within the population and using enriched breads and other potentially widely consumed food products as the benchmark for enrichment of other products can facilitate further acceptance consumers. Future research should assess the effects of derived arabinoxylans from selected cereal fibres (rye, sorghum, rice and corn) and energy crop sugarcane fibres on the rheological, sensory and physical effects in muffins production. Updated and accurate rheological properties food supplementation data is important for the food industry.

Author details

Roy Orain Porter
Guyana Food Safety Authority, Ministry of Agriculture, Georgetown, Guyana

*Address all correspondence to: star.troy@yahoo.com

References

[1] Alan, P. A., Ofelia, R. S., Patricia, T., & Maribel, R. S. R. (2012). Cereal bran and wholegrain as a source of dietary fibre: technological and health aspects. *International Journal of Food Sciences and Nutrition, 63*(7), 882-892. https://doi.org/10.3109/09637486.2012.676030

[2] Aydogdu, A., Sumnu, G., & Sahin, S. (2018). Effects of addition of different fibers on rheological characteristics of cake batter and quality of cakes. *Journal of food science and technology, 55*(2), 667-677. https://doi.org/10.1007/s13197-017-2976-y

[3] Dhingra, D., Michael, M., Rajput, H., & Patil, R. T. (2012). Dietary fibre in foods: a review. *Journal of Food Science and Technology, 49*(3), 255-266. https://doi.org/10.1007/s13197-011-0365-5

[4] Ekpa, O., Palacios-Rojas, N., Kruseman, G., Fogliano, V., & Linnemann, A. R. (2019). Sub-Saharan African Maize-Based Foods - Processing Practices, Challenges and Opportunities. *Food Reviews International, 35*(7), 609-639. https://doi.org/10.1080/87559129.2019.1588290

[5] Chen, Z., Li, S., Fu, Y., Li, C., Chen, D., & Chen, H. (2019). Arabinoxylan structural characteristics, interaction with gut microbiota and potential health functions. *Journal of Functional Foods, 54*, 536-551. https://doi.org/10.1016/j.jff.2019.02.007

[6] European Food Safety Authority Panel on Dietetic Products, N., & Allergies. (2010). Scientific Opinion on Dietary Reference Values for carbohydrates and dietary fibre. EFSA Journal, 8(3), 1462. https://doi.org/10.2903/j.efsa.2010.1462

[7] Li, W., Hu, H., Wang, Q., & Brennan, C. S. (2013). Molecular Features of Wheat Endosperm Arabinoxylan Inclusion in Functional Bread. *Foods (Basel, Switzerland), 2*(2), 225-237. https://doi.org/10.3390/foods2020225

[8] Scientific Advisory Committee on Nutrition (2015). *SACN carbohydrates and health report*. Retrieved from https://www.gov.uk/government/publications/sacn-carbohydrates-and-health-report

[9] Buttriss, J. L. (2015). Fibre - Need to increase intake according to new recommendations. *Nutrition Bulletin, 40* (4), 291-295. https://doi.org/10.1111/nbu.12174

[10] European Commission. (2020). *Dietary fibres*. European Commission. Retrieved September 1, 2020 from https://ec.europa.eu/jrc/en/health-knowledge-gateway/promotion-prevention/nutrition/fibre

[11] World Health Organisation. (2020). The Ottawa Charter for Health Promotion. World Health Organisation. Retrieved September 1, 2020 from https://www.who.int/health promotion/conferences/previous/ottawa/en/index1.html

[12] Shahbandeh, M. (2019). Size of the global dietary fibers market from 2018 to 2025. Statista. Retrieved August 18, 2020 from https://www.statista.com/statistics/1053431/dietary-fibers-market-value-global

[13] Foschia, M., Peressini, D., Sensidoni, A., & Brennan, C. S. (2013). The effects of dietary fibre addition on the quality of common cereal products. *Journal of Cereal Science, 58*(2), 216-227. https://doi.org/10.1016/j.jcs.2013.05.010

[14] Casas, G. A., Lærke, H. N., Bach Knudsen, K. E., & Stein, H. H. (2019). Arabinoxylan is the main polysaccharide in fiber from rice

coproducts, and increased concentration of fiber decreases in vitro digestibility of dry matter. *Animal Feed Science and Technology, 247*, 255-261. https://doi.org/10.1016/j.anifeedsci.2018.11.017

[15] Fadel, A., Mahmoud, A. M., Ashworth, J. J., Li, W., Ng, Y. L., & Plunkett, A. (2018). Health-related effects and improving extractability of cereal arabinoxylans. *International journal of biological macromolecules, 109*, 819-831. https://doi.org/10.1016/j.ijbiomac.2017.11.055

[16] Dimopoulou, M., & Kontogiorgos, V. (2020). Soluble dietary fibres from sugarcane bagasse. *International Journal of Food Science & Technology, 55*(5), 1943-1949. https://doi.org/10.1111/ijfs.14445

[17] Haghighi-Manesh, S., & Azizi, M. H. (2018). Integrated extrusion-enzymatic treatment of corn bran for production of functional cake. *Food science & nutrition, 6*(7), 1870-1878. https://doi.org/10.1002/fsn3.738

[18] Yadav, M. P., Kale, M. S., Hicks, K. B., & Hanah, K. (2017). Isolation, characterization and the functional properties of cellulosic arabinoxylan fiber isolated from agricultural processing by-products, agricultural residues and energy crops. *Food Hydrocolloids, 63*, 545-551. https://doi.org/10.1016/j.foodhyd.2016.09.022

[19] Bultum, L. E., Emire, S. A., & Wolde, Y. T. (2020). Influence of full fat rice bran from Ethiopian rice milling industries on nutritional qualities, physicochemical and sensory properties of bread and biscuits. *Journal of Food Measurement and Characterization, 14*(4), 2253-2261. https://doi.org/10.1007/s11694-020-00472-7

[20] Javaria, S. (2017). Formulation and quality evaluation of aglutenics biscuits supplimented with rice bran for coeliac patients. *Pure and Applied Biology, 6*. https://doi.org/10.19045/bspab.2017.600137

[21] Mena, B., Fang, Z., Ashman, H., Hutchings, S., Ha, M., Shand, P. J., & Warner, R. D. (2020). Influence of cooking method, fat content and food additives on physicochemical and nutritional properties of beef meatballs fortified with sugarcane fibre. *International Journal of Food Science & Technology, 55*(6), 2381-2390. https://doi.org/10.1111/ijfs.14482

[22] Rosicka-Kaczmarek, J., Komisarczyk, A., Nebesny, E., & Makowski, B. (2016). The influence of arabinoxylans on the quality of grain industry products. *European Food Research and Technology, 242*(3), 295-303. https://doi.org/10.1007/s00217-015-2549-0

[23] Singh, M., Liu, S. X., & Vaughn, S. F. (2012). Effect of corn bran as dietary fiber addition on baking and sensory quality. *Biocatalysis and Agricultural Biotechnology, 1*(4), 348-352. https://doi.org/10.1016/j.bcab.2012.02.005

[24] Jarosław Wyrwisz, M. K. (2015). The Application of Dietary Fiber in Bread Products. Journal of Food Processing & Technology, 06(05). https://doi.org/10.4172/2157-7110.1000447

[25] Ozturk, O. K., & Mert, B. (2018). The effects of microfluidization on rheological and textural properties of gluten-free corn breads. *Food Research International, 105*, 782-792. https://doi.org/10.1016/j.foodres.2017.12.008

[26] Sairam, S., Krishna, A. G. G., & Urooj, A. (2011). Physico-chemical characteristics of defatted rice bran and its utilization in a bakery product. *Journal of Food Science and Technology-Mysore, 48*(4), 478-483. https://doi.org/10.1007/s13197-011-0262-y

[27] Mendis, M., Leclerc, E., & Simsek, S. (2016). Arabinoxylans, gut

microbiota and immunity. *Carbohydrate polymers, 139*, 159-166. https://doi.org/10.1016/j.carbpol.2015.11.068

[28] Park, H.-Y., Lee, K.-W., & Choi, H.-D. (2017). Rice bran constituents: immunomodulatory and therapeutic activities. *Food & Function, 8*(3), 935-943. https://doi.org/10.1039/c6fo01763k

[29] Phimolsiripol, Y., Mukprasirt, A., & Schoenlechner, R. (2012). Quality improvement of rice-based gluten-free bread using different dietary fibre fractions of rice bran. *Journal of Cereal Science, 56*(2), 389-395. https://doi.org/10.1016/j.jcs.2012.06.001

[30] Bauer, J. L., Harbaum-Piayda, B., Stöckmann, H., & Schwarz, K. (2013). Antioxidant activities of corn fiber and wheat bran and derived extracts. LWT - Food Science and Technology, 50(1), 132-138. https://doi.org/10.1016/j.lwt.2012.06.012

[31] Begum, R., Ahmed, S., Hakim, M., & Sen, J. (2019). Comparative Study among Composite Breads Incorporated with Full Fatted and Defatted Rice Bran. *Journal of Environmental Science and Natural Resources, 11*(1-2), 43-52. https://doi.org/10.3329/jesnr.v11i1-2.43364

[32] Gul, K., Yousuf, B., Singh, A. K., Singh, P., & Wani, A. A. (2015). Rice bran: Nutritional values and its emerging potential for development of functional food—A review. *Bioactive Carbohydrates and Dietary Fibre, 6*(1), 24-30. https://doi.org/10.1016/j.bcdf.2015.06.002

[33] Hussein, A., & Ibrahim, G. (2019). Effects of various brans on quality and volatile compounds of bread. *Foods and Raw Materials, 7*, 35-41. https://doi.org/10.21603/2308-4057-2019-1-42-50

[34] Rose, D. J., Inglett, G. E., & Liu, S. X. (2010). Utilisation of corn (*Zea mays*) bran and corn fiber in the production of food components. *Journal of the Science of Food and Agriculture, 90*(6), 915-924. https://doi.org/10.1002/jsfa.3915

[35] Koegelenberg, D., & Chimphango, A. F. A. (2017). Effects of wheat-bran arabinoxylan as partial flour replacer on bread properties. *Food Chemistry, 221*, 1606-1613. https://doi.org/10.1016/j.foodchem.2016.10.130

[36] Sandhu, P., Bains, K., Singla, G., & Sangwan, R. (2018). Utilization of Corn Fibre and Pectin Gel for the Development of Low Calorie and High Fibre Biscuits. *Current Journal of Applied Science and Technology, 25*(5), 1-9. https://doi.org/10.9734/cjast/2017/38966

[37] Sangeetha, A. V., Mahadevamma, S., Begum, K., & Sudha, M. L. (2011). Influence of processed sugarcane bagasse on the microbial, nutritional, rheological and quality characteristics of biscuits. *International Journal of Food Sciences and Nutrition, 62*(5), 457-464. https://doi.org/10.3109/09637486.2010.549819

[38] Vijerathna, M. P. G., Wijesekara, I., Perera, R., Maralanda, S. M. T. A., Jayasinghe, M., & Wickramasinghe, I. (2019). Physico-chemical Characterization of Cookies Supplemented with Sugarcane Bagasse Fibres. *Vidyodaya Journal of Science, 22* (1), 29. https://doi.org/10.4038/vjs.v22i1.6062

[39] Yadav, R. B., Yadav, B. S., & Chaudhary, D. (2011). Extraction, characterization and utilization of rice bran protein concentrate for biscuit making. *British Food Journal, 113*(9), 1173-1182. https://doi.org/10.1108/00070701111174596

[40] Zhang, L., Van Boven, A., Mulder, J., Grandia, J., Chen, X. D., Boom, R. M., & Schutyser, M. A. I. (2019). Arabinoxylans-enriched fractions: From

dry fractionation of wheat bran to the investigation on bread baking performance. *Journal of Cereal Science, 87*, 1-8. https://doi.org/10.1016/j.jcs.2019.02.005

[41] Moher D, Liberati A, Tetzlaff J, Altman DG, The PRISMA Group (2009). Preferred reporting items for systematic reviews and MetaAnalyses: The PRISMA statement. *PLoS Med* 6(7): e1000097. https://doi.org/10.1371/journal.pmed1000097

[42] Downs, S. H., & Black, N. (1998). The feasibility of creating a checklist for the assessment of the methodological quality both of randomised and non-randomised studies of health care interventions. *Journal of epidemiology and community health, 52*(6), 377–384. https://doi.org/10.1136/jech.52.6.377

[43] Hooper, P., Jutai, J. W., Strong, G., & Russell-Minda, E. (2008). Age-related macular degeneration and low-vision rehabilitation: a systematic review. Canadian Journal of Ophthalmology, 43(2), 180-187. https://doi.org/10.3129/i08-001

[44] Gil-López, D. I. L., Lois-Correa, J. A., Sánchez-Pardo, M. E., Domínguez-Crespo, M. A., Torres-Huerta, A. M., Rodríguez-Salazar, A. E., & Orta-Guzmán, V. N. (2019). Production of dietary fibers from sugarcane bagasse and sugarcane tops using microwave-assisted alkaline treatments. *Industrial Crops and Products, 135*, 159-169. https://doi.org/10.1016/j.indcrop.2019.04.042

[45] Amna, Y., Bhatti, M. S., Anwaar, A., & Randhawa, M. A. (2011). Effect of rice bran supplementation on cookie baking quality. *Pakistan Journal of Agricultural Sciences, 48*(2), 133-138.

[46] Jauharah, M. Z., Wan Ishak, W. R., & Robert, S. (2014). Physicochemical and Sensorial Evaluation of Biscuit and Muffin Incorporated with Young Corn Powder. Sains Malaysiana, 43, 45-52.

[47] Ameh, M. O., Gernah, D. I., & Igbabul, B. D. (2013). Physico-Chemical and Sensory Evaluation of Wheat Bread Supplemented with Stabilized Undefatted Rice Bran. Food and Nutrition Sciences, 04(09), 43-48. https://doi.org/10.4236/fns.2013.49a2007

[48] Pauline, M., Roger, P., Sophie Natacha Nina, N. E., Arielle, T., Eugene, E. E., & Robert, N. (2020). Physico-chemical and nutritional characterization of cereals brans enriched breads. *Scientific African, 7*, e00251. https://doi.org/10.1016/j.sciaf.2019.e00251

[49] Silva, A. S., Correa, L. G., Kanai, R. S. S., & Shirai, M. A. (2020). Effect of sugarcane bagasse addition on physical, chemical, and sensory properties of oat flour and banana cake. *Journal of Texture Studies*. https://doi.org/10.1111/jtxs.12542

[50] European Food Safety Authority Panel on Dietetic Products, N., & Allergies. (2011). Scientific Opinion on the substantiation of health claims related to arabinoxylan produced from wheat endosperm and reduction of post-prandial glycaemic responses (ID 830) pursuant to Article 13(1) of Regulation (EC) No 1924/2006. EFSA Journal, 9(6), 2205. https://doi.org/10.2903/j.efsa.2011.2205

[51] Saeed, F., Pasha, I., Anjum, F. M., & Sultan, M. T. (2011). Arabinoxylans and Arabinogalactans: A Comprehensive Treatise. Critical Reviews in Food Science and Nutrition, 51(5), 467-476. https://doi.org/10.1080/10408391003681418

[52] Ortiz de Erive, M., Wang, T., He, F., & Chen, G. (2020). Development of high-fiber wheat bread using microfluidized corn bran. Food Chemistry, 310, 125921. https://doi.org/10.1016/j.foodchem.2019.125921

[53] Leang, Y. H., & Saw, H. Y. (2011). Proximate and functional properties of

sugarcane bagasse. *Agro Food Industry Hi-Tech, 22*(2), 5-8.

[54] Kale, M. S., Pai, D. A., Hamaker, B. R., & Campanella, O. H. (2010). Structure-function relationships for corn bran arabinoxylans. *Journal of Cereal Science, 52*(3), 368-372. https://doi.org/10.1016/j.jcs.2010.06.010

[55] Zidan, N., & Eldemery, M. (2016). Utilization of defatted black rice bran in wheat bread preparation for enhancing nuritional and functional properties. *Journal of Food and Dairy Sciences, 7*(2), 107-117. https://doi.org/10.21608/jfds.2016.42819

[56] Sozer, N., Cicerelli, L., Heiniö, R.-L., & Poutanen, K. (2014). Effect of wheat bran addition on in vitro starch digestibility, physico-mechanical and sensory properties of biscuits. *Journal of Cereal Science, 60*(1), 105-113. https://doi.org/10.1016/j.jcs.2014.01.022

[57] Mishra, N., & Chandra, R. (2012). Development of functional biscuit from soy flour & rice bran. *International Journal of Agricultural and Food Science, 2*, 14-20.

[58] Jia, M., Yu, Q., Chen, J., He, Z., Chen, Y., Xie, J., . . . Xie, M. (2020). Physical quality and in vitro starch digestibility of biscuits as affected by addition of soluble dietary fiber from defatted rice bran. *Food Hydrocolloids, 99*, 105349. https://doi.org/10.1016/j.foodhyd.2019.105349

[59] Döring, C., Nuber, C., Stukenborg, F., Jekle, M., & Becker, T. (2015). Impact of arabinoxylan addition on protein microstructure formation in wheat and rye dough. *Journal of Food Engineering, 154*, 10-16. 10.1016/j.jfoodeng.2014.12.019

[60] El-Gammal, R., & El Kewawy, H. (2014). Effect of addition of stabilized rice bran on physical, rheological and chemical characteristics of pan bread. *Journal of Food and Dairy Sciences, 5*(11), 813-825. https://doi.org/10.21608/jfds.2014.53231

[61] Lockyer, S., & Spiro, A. (2020). The role of bread in the UK diet: An update. *Nutrition Bulletin, 45*(2), 133-164. https://doi.org/10.1111/nbu.12435

[62] Mozaffarian, D. (2016). Dietary and Policy Priorities for Cardiovascular Disease, Diabetes, and Obesity. *Circulation, 133*(2), 187-225. https://doi.org/10.1161/circulationaha.115.018585

[63] Ayala-Soto, F. E., Serna-Saldívar, S. O., & Welti-Chanes, J. (2017). Effect of arabinoxylans and laccase on batter rheology and quality of yeast-leavened gluten-free breads. *Journal of Cereal Science, 73*, 10-17. https://doi.org/10.1016/j.jcs.2016.11.003

Chapter 3

Honey Production Process

Emek Dümen, Nadide Gizem Tarakçı and Gözde Ekici

Abstract

Honey has been considered as a very important and superior nutrient in human nutrition since ancient times due to its ability to be consumed by humans without processing, easy digestibility, nutritional properties and biological benefits. Although honey contains many desired bioactive and antibacterial substances, which may be sufficient for antimicrobial activity, it cannot be produced in sufficient quantities due to low water activity under normal conditions. This causes various food and bee-borne spores/non-spores pathogens going viral. Hence, it may cause the risk of parasitological and fungal agents to be found. In honey production, "Hazard Analysis Critical Control Point (HACCP)" must be applied meticulously and completely. Current technologies in honey production will be explained in this section.

Keywords: honey production, microbiological risks, HACCP

1. Introduction

Honey consumption has a very old history for humans. It has been used as a sweetener and flavoring in countless food and beverages. Since ancient times, honey has been known for its nutritious and therapeutic aspects. The most important components of honey are carbohydrates, which are found in the form of fructose, glucose, disaccharides, and oligosaccharides, and components such as maltose, isomaltose, maltulose, sucrose provide the sweet taste to honey. It also contains enzymes such as amylase, oxidase peroxide, catalase and acid phosphorylase, including anderosis and panoz. Also, honey is rich in amino acids, minerals, antioxidants and various phytochemicals [1]. Many of the reported biological properties of honey, such as antioxidant, antibacterial, antifungicidal, anti-inflammatory, hypotensive, antiproliferative, hepato-protective properties of these components are associated with presence of these properties. However, the composition of honey largely depends on a number of factors, such as flower source, geographical region, climatic conditions, harvesting season, processing and storage conditions. There are studies that report that honey, administered alone or in combination with traditional therapy, may be useful in the treatment of chronic diseases that are commonly associated with oxidative stress and the state of inflammation [2]. Honey is classified according to various criteria. In this classification, honey is classified as secretion honey (such as pine honey, oak honey, fir honey, leaf honey) and flower honey (linden honey, cotton honey, trirose honey, thyme honey, mashed honey, acacia honey, heather honey, etc.). According to the form of marketing, framed comb honey, natural comb honey, partial comb honey, cut-comb honey, strained honey, crystallized honey, creamed honey, pressed honey, chunk honey (strained with comb or comb with strained), filtered honey and baker's honey. According to the moisture content, honey is classified as grade 1 honeys (humidity below 17.8%), grade 2 honeys (humidity up to 18.6%) and grade 3 honeys (humidity up to 20%). According to their color, honey is classified as white, golden, amber and dark. The color of honey can vary from light water white to black warehouse [3]. The physical and chemical properties, antimicrobial effects, which are of great importance for public health, and GMP and HACCP systems applied in the production process and microbiological dangers will be addressed in this section.

1.1 Physical and chemical properties of honey

Honey contains about 200 substances and is a nutrient consisting of substances such as carbohydrates, water, enzymes, free amino acids, essential minerals, vitamins, phenolic compounds, volatile compounds (monoterpenes, benzene derivatives) and some other solids. Carbohydrates in honey are mainly monosaccharides, glucose and fructose. This is followed by disaccharides and trisaccharides. They contribute mainly to the energy value. Proteins include enzymes such as invertase, diastase, glucose oxidase, catalase, peroxidase and acid phosphatase, and their content varies from 0.1% to 3.3% depending on the type of honey. It contains essential and non-essential amino acids, but the most common amino acid in honey is proline, which accounts for 1% of honey components [2]. Honey contains tocopherol (E), anti-hemorrhagic vitamin (K), ascorbic acid (C), thiamine (B_1), riboflavin (B_2), niacin (B_3), pantothenic acid (B_5) and a small amount of vitamin pyridoxine (B_6). Vitamins of the B complex and vitamin C are mainly derived from pollen and can be affected by commercial and industrial processes such as filtration or oxidation reactions [4].

Honey has a slight acid reaction due to about 0.57% organic acids. Acids contribute to the aroma and antimicrobial activity of honey. The predominant acid in honey is gluconic acid, it is followed by aspartic citric, acetic, formic, fumaric, galacturonic, malonic, formic, acetoglutaric, glutamic, butyric, glutaric, propionic, pyruvic, glioxia, 2-hydroxybutyric, a-hydroxiglutaric, isocyric, lactic, malic, methylmalonic, kynic, succinic, tartaric, oxalic acid [2]. The mineral content in honey ranges from 0.04% in light honey and 0.2% in dark honey. Potassium is the most abundant element. But the main bioactive molecules contained in honey are represented by polyphenols. Polyphenols are a heterogeneous chemical compound that can be divided into flavonoids (flavonols, flavones, flavanols, flavanones, anthocyanin, calcones and isoflavones) and non-flavonoid (phenolic acids). The profile of polyphenolic compounds in honey is thoroughly studied and includes vanillin, caffeic, syringing, p-gamic, ferulic, ellagic, 3-hydroxybenzoic, chlorogenic, genistic, gallic and benzoic acids and contains different phenolic acids, such as different flavonoids, mainly quercetin, kaempferol, myricetin, chrysin, galangin, hesperetin. The amount and type of polyphenols largely depends on the flower source or the variety of honey. In addition, it is known that there is a strong relationship between antioxidant activity and total phenolic content [5].

1.2 The importance of honey in terms of health and its antimicrobial effect

Honey is a food that has been used in therapeutic treatments for thousands of years. Among other useful properties to health, this product has been reported as a promising agent for wound healing, including leg ulcers and eyes, skin disorders by in vitro and clinical studies. In studies on New Zealand manuka honey, unique to the New Zealand, positive effects were observed on the viability of potentially useful *Lactobacillus reuter* and *Bifidobacterium longum* found in the human intestine. Moreover, it was found that *Salmonella enterica* Typhimurium, an enteric pathogenic bacterial type, showed a 65% reduction in their proliferation. In this sense, it has been established that manuka honey has a beneficial effect on the intestine by producing acid metabolites that reduce the intestinal pH and prevent pathogenic colonization and hence support the growth of bifidobacteria and lactic acid bacteria. Honey has been reportedly able to modulate oxidative stress and also has anti-proliferative, pro-apoptotic, anti-inflammatory and anti-metastatic properties. The anticancer effect of honey is connected to the presence of natural bioactive compounds, mainly such as pinobanksin, pinocembrin, luteolin, chrysin, salicylic acid and 3.4 dihydroxybenzoic acid [6, 7]. Some of the vitamins contained in honey are ascorbic acid, pantothenic acid, niacin and riboflavin. Moreover, it is a food that also contains minerals such as calcium, copper, iron, magnesium, manganese, phosphorus, potassium and zinc. Its rich variety of vitamins and minerals also plays a role in increasing the antioxidant characteristics of honey. The presence of free radicals and reactive oxygen types is responsible for pathogenesis of aging, as well as cellular dysfunction, metabolic and cardiovascular diseases. Consumption of foods rich in antioxidants can protect against these pathological changes, preventing the pathogenesis of chronic ailments [8].

Various parameters such as low water activity, high sugar content, acidity and hydrogen peroxide (H_2O_2) content, phytochemicals, peptides, non-peroxidase glycoeptides and proteins make up the antibacterial potential of honey. Water activity of honey varies from 0.56–0.62. These values might be considered low enough to prevent the development of bacteria or other microorganisms [9]. Although previously it was believed that the only responsible agent for the antibacterial effect of diluted honey was H_2O_2 and that this antibacterial effect can be completely eliminated through catalysis, it has been found out that bacteria can also be affected via the existence of pythochemical elements present in honey [10]. As it suppresses the activities of bacteria causing infections in urinary systems, such as *E. coli* and *Proteus* species and *Streptococcus faecalis*, diluted honey is used to treat urinary system infections and it inhibits toxin production [11]. Undiluted honey hinders the reproduction and development of bacteria due to the content of sugar, which exerts osmotic pressure on bacterial cells and causes water to flow out of bacterial cells through osmosis. Thus, the cells shrink due to dehydration, and they cannot remain alive in hypertonic sugar solution. The optimal pH necessary for the development of most microorganisms ranges from 6.5–7.5. The pH value of honey is between 3.2–4.5, and this value is a very distinctive feature of its antibacterial activity. This acidity is caused by the presence of certain important organic acids, especially gluconic acid - in 0.5% (a/h) concentration. Glycogenic acid is produced from glucose oxidation by an endogenous enzyme of glucose oxidase and is an extremely powerful antibacterial agent. In undiluted pure honey, low pH can contribute to antibacterial action, but when the product is diluted pH alone is not enough to prevent the development of bacteria [9]. The formation of H_2O_2 is a dominant mechanism in which honey exerts bacteriostatic and bactericidal activity. It provides antibacterial activity of honey and is produced enzymatically. The enzyme glucose oxidase is inherently inactive in honey due to low pH conditions, and glucose oxidase is activated when honey is diluted. However, it is known that concentrations of H_2O_2 are adversely affected by various minor components, such as nectar, pollen, and yeast. It has also been reported as having high sensitivity to light and light sources [12]. Honey contains relatively small amounts of proteins, whose molecular weights range from 20 to 80 kDa, ranging from approximately 0.1% to 0.5%. These proteins contain many enzymes involved in sugar metabolism, such as alpha and beta glucosidase, glucose oxidase and amylase. Numerous studies have shown that important royal jelly proteins have antimicrobial and anticancer activity and anti-inflammatory potential [13].

Honey shows antibacterial activity against a large number of bacteria in different environments. Natural components of honey have antifungal, antiviral, antibacterial activities. It has been reported that the antibacterial activity of honey is also likely to depend on the pasture, climatic conditions, and also on the natural composition of flower nectar. Honey has excellent antibacterial activity against methicillin-resistant *Staphylococcus aureus* (MRSA), often associated with wound and burn infections, and *Pseudomonas* spp. Many studies have shown that honey is also effective against hemolytic streptococci and vancomycin resistant enterococci. Twenty-one kinds of honey tested for antibacterial activity against *Staphylococcus aureus (S. aureus)* and *Pseudomonas aeruginosa (P. aeruginosa)*, and it has been established that they have a positive effect due to H_2O_2 and polyphenolic content levels. The effectiveness of free radical cleansing is observed in all kinds of honey. In addition, honey tested by freezing, drying and powdering has been reported to show antioxidant activity in each form [9, 14]. Flavonoids contained in the natural composition of honey is also known to be effective against microorganisms that are present in the tissue of chronic wounds, in particular *S. aureus*, *P. aeruginosa*, as well as *Escherichia coli (E. coli)*. Flavonoids are often recommended as a natural source to control chronic inflammatory diseases, the incidence of which increases significantly. Despite the fact that the topical application of honey for medicinal purposes is old, there are a small number of studies that address its anti-inflammatory activity at the cellular level. Although flavonoids are small components of honey, their anti-inflammatory effect is extraordinary compared to other natural compounds [15]. Honey was found to have a preventive effect on about 60 bacteria such as, *Bacillus anthracis, Corynebacterium diptheriae, Haemophilus influenzae, Klebsiella pneumoniae, Shigella, Mycobacterium tuberculosis*, and many aerob and anaerob bacterial types. In vitro studies of *Helicobacter pylori* in the human digestive system have shown that when using honey, its activity decreases by 20%. It has been reported that honey can be used in combination with antibiotics to produce a synergistic effect of bactericidal activity against *Helicobacter pylori*. The main difference of honey with antibiotics is that it does not develop antibiotic-resistant bacteria, so it can be used continuously without such risk [16].

2. Other beekeeping products

2.1 Pollen

Pollen is the only source of protein found in nature for bees. The amino acids contained in its composition are isolosin, arginine, lysine, histidine, leucine, methionine, treonine, phenylalanine, tryptophan and valine. It is essential for adequate development of their muscles, tissues, secretory glands and other organs in the upbringing of honeybees and its young stages. It is a nutritional source rich in vitamins, proteins, sterols, minerals and lipids. It has been reported that pollen collected by honeybees may have differences in their general chemical composition as a result of supplying from different plants [17].

2.2 Nectar

Bees have two stomach and they use one of them to perform normal body functions whereas the other to store the nectar they collect. In order to collect nectar found in flowers, bees use rod-like, tubular long tongues. It has been reported that bees can contain about 70 mg of nectar in the stomach they store nectar, and that they should visit 100 to 1500 different flowers to fully fill their honey stomach [17].

2.3 Propolis

Propolis is recognized as a therapeutic agent due to several reported functional effectiveness. It is known that honey contains phenolic compounds. Propolis contains a higher content of phenolic compounds than honey and shows significantly higher antimicrobial and antioxidant activities. Today it is used in industry as a component of confectionery, biopharmaceuticals and cosmetics. It is gaining popularity as a natural preservative and helps to improve shelf life and consumer health as a source of bioactive compounds for food and drinks. However, propolis has a strong and bitter taste, which changes the sensory properties of food due to the high concentration of phenolic compounds. Therefore, the acceptance of foods containing propolis by consumers must be determined by its propolis concentration, which has to be carefully researched so as not to adversely change the sensory properties of such foods [18].

2.4 Bee milk

The importance of bee milk, one of bee products, was noticed in the 1600s and was given the name "*Royal Jelly*", which means excellent food in English. Bee milk is secreted from the upper jaw (mandibular) and throat glands (hypopharyngeal) of young worker bees of 5–15 days of age. All larvae only in their first three-day period, and the larvae that will become the queen bee are fed with royal jelly during the entire larval and adult periods. Bee milk can be described as food with a peculiar smell and a bitter taste and a mush-like form with bone-like color. It is collected from the cells of larvae of 3–4 days of age of the future queen, or from the cells of queen bees where larvae of 1–2 days are laid after 48–72 hours. It is quite flowing and has yogurt-like consistency but is a homogeneous substance. It has a light beige and yellowish whitish color, a sharp phenolic smell and a distinctive sour taste. Its density is approximately 1.1 g/cm^3 and is soluble in water [19].

3. Microbiological risks in honey and honey products

Although honey is considered a low-risk food due to its antimicrobial and bacteriostatic effects, studies disprove this view. In addition to primary contamination, staff, tools and equipment used in beekeeping and honey production are also a potential source of secondary contamination. In addition, honey, which has the potential to contain many microorganisms as a result of cross-contamination, is among the important nutrients and can indirectly threaten public health. Despite the fact that some types of honey contain H_2O_2 and benzoic acid and phenolic compounds such as some flavonoids, it can constitute risks for consumer health due to minimal hygiene rules. It is reported that pathogens can be found as causative agents in honey produced without food safety systems. Food-borne pathogens are recognized as an important risk factor for public health in developed and developing countries due to their spread around the world. Viruses, bacteria, fungi, parasites and mites are the most common disease factors in beekeeping. Fecal-oral route is an important way of transmission of these diseases. Agents that pollute bees through water and food can be transmitted to larvae by infected bees. Another contamination that may occur in honey is secondary contamination caused by secondary contamination sources such as personnel, tools and equipment [20].

The presence of strains *Bettsya alvei*, *Acosphaera apis* and *Acosphaera* major in honey production can be indicative of improper beehive management practices. Different types of microorganisms such as *Acinetobacter* spp., *Bacillus* spp.,

Clostridium spp., *Corynebacterium* spp., *Pseudomonas* spp. are bacteria that are widely found in the soil. *Brochothrix* spp., *Citrobacter* spp., *Enterobacter* spp., *Erwinia* spp., *Flavobacterium* spp., *Lactobacillus* spp., *Lactococcus* spp., *Leuconostoc* spp., *Listeria* spp. and *Pediococcus* spp. are other bacteria that are likely to be found in plants and plant products. On the other hand, among yeast strains *Saccharomyces*, *Schizosaccharomyces* and *Torula* species predominate in high humidity sugars. Bacterial spores, especially *Bacillus* and *Clostridium*, can be seen in honey. *Clostridium* is an indicator organism that provides evidence of contamination or pollution in honey. *Clostridium botulinum* (*C. botulinum*) spores are usually found at low levels in honey. The presence of *clostridium* spores can be dangerous, especially for children under one year of age. It is known that infant botulism is mainly caused by the consumption of honey contaminated with *C. botulinum* [21, 22]. *C. botulinum* forms 4 different types of neuroparalytic diseases in humans. In addition to infant botulism, they are classified as food-borne botulism, wound botulism, and yet unclassified latent botulism. The most important of them is infant botulism, which occurs in newborn babies of 3–20 weeks. Infant botulism is diagnosed with isolation of *C. botulinum* and toxin in feces. Decreases in sucking and swallowing reflexes of infants can be observed, which is very rarely fatal [23].

One of the animal products that have been the focus of food warnings due to the presence of chemical hazards such as antibiotics or pesticides in recent years are honey and honey products. The source of these residues in honey is mainly due to bee parasites, such as European offspring rot (*Streptococcus pluton*) or American offspring rot (*Bacillus larvae*) and are veterinary drugs that are necessary to treat bacterial diseases. It is known that chemical residues caused by these drugs used to eliminate microbiological risks, lead to such adverse conditions on human health as allergic reactions, bacterial resistance, along with changes of reproductive toxicity [24, 25].

4. HACCP in honey production

Food safety can be ensured by systematic implementation of all activities in line with a plan. The Hazard Analysis Critical Control Points (HACCP) system, as a preventive system for ensuring food safety, controls production at various points throughout the food production, thereby ensuring that the final product complies with legislation. Preliminary Requirement Program must be created first in order to establish the HACCP system in any food business. In this context, the deficiencies of the infrastructure and processes such as water, energy, warehouse, cleaning and sanitation, personnel, environment and equipment hygiene, personnel training and pest control should be addressed. However, it is necessary to plan the process management by writing down the procedures. The processes that need to be addressed afterwards can be sorted as follows; identification of the HACCP team and a clear definition of the task descriptions by making the managerial organization chart, determination of food safety policy by business management, making an understandable description of the products to be produced, determination of the intended usage method, preparation of a flow diagram and placement plan by HACCP team and verification of this plan at site, analyzing hazards and risks, identification of critical control points, making up of critical limits and monitoring procedures, determination of corrective activities for cases where it is necessary, and the proving or verification of the effectiveness of the system [26].

Codex Alimentarius Standard and the European Commission allows nomenclature for honeys produced from certain botanical sources if the product comes from the specified origin and has anticipated physicochemical, organoleptic and microscopic properties. The fact that there are different varieties of honey and each has its own production steps, leads to an increase in the limits that need to be controlled. For the import of food products of animal origin, such as honey, EU legislation requires a number of health and national residue monitoring procedures such as HACCP during the production and processing of honey. These requirements are known to be independent of whether honey is organic or traditional. Thus, imported products are intended to meet the standards required for production and trade within EU member countries. Costs, lack of qualified personnel, misinterpretation of EU legislation, lack of laboratory in international standards and improper infrastructure are the main obstacles to being accredited by the EU [27, 28].

Although honey is a product that is part of the low-risk group due to its high sugar content, it should be carefully examined for physical, chemical and biological hazards. In general, the hygiene of the processing area, tool and equipment and personnel in contact with food should be observed as it should be in all food enterprises. Physical hazards such as soil, plant materials, glass materials, tools and equipment are defined as potential hazards to honey. Traces of pesticides and herbicide, beekeeping drugs and antibiotics are chemical hazards. Soil originated *C. botulinum*, the most important biological danger in honey production, is eliminated by the provision of hygienic conditions in the production of honey [29].

General hygiene rules should be applied effectively to prevent physical, chemical, and microbiological hazards. In hives, legally approved preservatives should be used. Insects and mice should be kept away from hives. During transportation, the vehicles should be cleaned well, in case of the presence of dirt left from the previous use. It is necessary to effectively clean the equipment and work area before and after use. Persons involved in the process should wear a separate clothing to protect the product from contamination caused by clothing or individuals. Especially before and after use, cleaning control of filters must be carried out effectively. In the HACCP plan, the purpose of conducting hazard analysis must be effectively controlled. All potential hazards in each step of the workflow process must be identified and the risk and severity of each identified hazard should be assessed. At this point, it is also necessary to determine the sources of dangers. In a study, corrective activities to prevent and/or eliminate hazards were determined and two critical controls points, "filtration/unloading" and "packaging" were pointed out. Examples of forms used in each HACCP plan and all procedures of the HACCP plan must be provided and monitored [26, 30].

5. Conclusions

HACCP system, which is successfully implemented in the food industry, is the most effective quality system in terms of the supply of safe products. The purpose of the use of HACCP system is to provide reliable food to the consumer with the desired characteristics and quality. Honey production, which is suitable to be affected by climatic conditions, should be made systematic and controllable by removing traditional methods that are difficult to trace. In this context, the creation of honey workflow process, determining potential hazards, and analyzing hazards, taking necessary precautions, recording the system, providing internationally reliable product guarantee is of great importance for public health as well as for the economies of countries. Effective implementation of the HACCP system in enterprises is inevitable so that retrospective monitoring and recall models can be used in the event of any negativity. In the production processes of foods with high nutritional value as honey, all necessary food safety requirements must be met to protect and improve public health.

Author details

Emek Dümen[1], Nadide Gizem Tarakçı[2] and Gözde Ekici[3*]

1 Food Hygiene and Technologies, Faculty of Veterinary, Istanbul University-Cerrahpasa, Istanbul, Turkey

2 Department of Nutrition and Dietetics, School of Health Sciences, Istanbul Medipol University, Istanbul, Turkey

3 Faculty of Health Sciences, Department of Nutrition and Dietetics, Istanbul Kultur University, Istanbul, Turkey

*Address all correspondence to: g.ekici@iku.edu.tr

References

[1] Meo SA, Al-Asiri SA, Mahesar AL, Ansari MJ. Role of honey in modern medicine. Saudi Journal of Biological Sciences. 2017;24:975-978. DOI: http://dx.doi.org/10.1016/j.sjbs.2016.12.010

[2] Terzo S, Mulè F, Amato A. Honey and obesity-related dysfunctions: a summary on health benefits. Journal of Nutritional Biochemistry. 2020;82:108401. DOI: https://doi.org/10.1016/j.jnutbio.2020.108401

[3] Doğan H. Determination of chemical, physical and antimicrobial properties of nectar honeys [thesis]. Atatürk University; 2014.

[4] Cianciosi D, Forbes-Hernández TY, Afrin S, Gasparrini M, Reboredo-Rodriguez P, Manna PP, Zhang J, Lamas LB, Florez SM, Toyos PA, Quiles JL, Giampieri F, Battino M. Phenolic compounds in honey and their associated health benefits: A review. Molecules. 2018; 23:2322. DOI:10.3390/molecules23092322

[5] Miguel MG, Antunes MD, Falerio ML. Honey as a Complementary Medicine. Integrative Medicine Insights. 2017; 12:1-15. DOI: https://doi.org/10.1177/117863371770286

[6] Seraglio SKT, Schulz M, Gonzaga VL, Fett R, Costa ACO. Current status of the gastrointestinal digestion effects on honey: A comprehensive review. Food Chemistry. 2021; 357: 129807. DOI: https://doi.org/10.1016/j.foodchem.2021.129807

[7] Afrin S, Haneefa SM, Fernandez-Cabezudo MJ, Giampieri F, Al-Ramadi BK, Battino M. Therapeutic and preventive properties of honey and its bioactive compounds in cancer: an evidence-based review. Nutrition Research Reviews. 2020;33(1):50-76. DOI: https://doi.org/10.1017/S0954422419000192

[8] Ajibola A, Chamunorwa JP, Erlwanger KH. Nutraceutical values of natural honey and its contribution to human health and wealth. Nutrition & Metabolism. 2012;9(1):1-12.

[9] Almasaudi S. The antibacterial activities of honey. Saudi Journal of Biological Sciences. 2021;28:2188-2196. DOI: https://doi.org/10.1016/j.sjbs.2020.10.017

[10] Albaridi NA. Antibacterial Potency of Honey. International Journal of Microbiology. 2019. DOI: https://doi.org/10.1155/2019/2464507

[11] Eteraf-Oskouei T, Najafi M. Traditional and Modern Uses of Natural Honey in Human Diseases: A Review. Iranian Journal of Basic Medical Sciences. 2013; 16(6): 731-742.

[12] Brudzynski K. A current perspective on hydrogen peroxide production in honey. A review. Food Chemistry. 2020;332:127229. DOI: https://doi.org/10.1016/j.foodchem.2020.127229

[13] Tonks AJ, Cooper RA, Jones KP, Blair S, Parton J, Tonks A. Honey stimulates inflammatory cytokine production from monocytes. Cytokine 2003;21:242-247. DOI: doi:10.1016/S1043-4666(03)00092-9

[14] Stagos D, Solitsiotis N, Tsadila C, Papaeconomou S, Arvanitis C, Ntontos A, Karkanta F, Adamou-Androulaki S, Petrotos K, Spandidos DA, Kouretas D, Mossialos D. Antibacterial and antioxidant activity of different types of honey derived from Mount Olympus in Greece. International Journal Of Molecular Medicine. 2018 42:726-734. DOI: 10.3892/ijmm.2018.3656

[15] Silva B, Biluca FC, Gonzaga LV, Fett R, Dalmarco EM, Caon T, Costa ACO. In vitro anti-inflammatory properties of honey flavonoids:

A review. Food Research International. 2021;14:110086. DOI: https://doi.org/10.1016/j.foodres.2020.110086

[16] Ilia G, Simulescu V, Merghes P, Varan N. The health benefits of honey as an energysource with antioxidant, antibacterial andantiseptic effects. Science & Sports. DOI: https://doi.org/10.1016/j.scispo.2020.10.005

[17] Sahin G. Research on the chemıcal composıtıon of some honey [thesis]. Ankara University; 2019.

[18] Osés SM, Pascual-Maté A, Fernández-Muiño MA, López-Díaz TM, Sancho MT. Bioactive properties of honey with propolis. Food Chemistry. 2016;196:1215-1223. http://dx.doi.org/10.1016/j.foodchem.2015.10.050

[19] Akyol E, Baran Y. Arı sütünün yapısı, insanlar ve arılar için önemi. Uludag Bee Journal. 2015; 15(1):16-21.

[20] Bayrakal GM, Ekici G, Akkaya H, Sezgin FH, Dümen E. Detection and Molecular Examination of Pathogens in Honey and Bees in the Northern Marmara Region, Turkey. Journal of Kafkas University of Veterinary Faculty. 2020;26(3):313-319. DOI: 10.9775/kvfd.2019.22845

[21] Finola MS, Lasagno MC, Marioli JM. Microbiological and chemical characterization of honeys from central Argentina. Food Chemistry. 2007;100:1649-1653. DOI:10.1016/j.foodchem.2005.12.046

[22] Vázquez-Quinones CR, Moreno-Terrazas R, Natividad-Bonifacio I, Quinones-Ramírez EI, Vázquez-Salinas C. Microbiological assessment of honey in México. Revista Argentina De Microbiologia. 2018;50(1):75-80. DOI: https://doi.org/10.1016/j.ram.2017.04.005

[23] Nevas M, Lindström M, Hörman A, Keto-Timonen R, Korkeala H. Contamination routes of *Clostridium botulinum* in the honey production environment. Environmental Microbiology. 2006;8(6):1085-1094. DOI:10.1111/j.1462-2920.2006.01000.x

[24] Juan-Borras M, Periche A, Domenech E, Escriche I. Routine quality control in honey packaging companies as a key to guarantee consumer safety. The case of the presence of sulfonamides analyzed with LC-MS-MS. Food Control. 2015;50:243-249. DOI: http://dx.doi.org/10.1016/j.foodcont.2014.08.021

[25] El-Nahhal Y. Pesticide residues in honey and their potential reproductive toxicity. Science of the Total Environment. 2020;741:139953. DOI: https://doi.org/10.1016/j.scitotenv.2020.139953

[26] Suna S, Özcan Sinir G, Anlar D. Bal Üretim Prosesinde HACCP Uygulaması. Arıcılık Araştırma Dergisi. Arıcılık Araştırma Dergisi. 2013;5(10):17-21.

[27] Bogdanov S, Fluri P, Imdorf A, Charrierre JD, Kilchenmann V. Self control sytstem for the production of high quality honey: The Swiss example. Apimondia Journal. 2005;40:28-33.

[28] Oddo LP, Bogdanov S. Determination of honey botanical origin: problems and issues. Apidologie. 2004;35:2-3. DOI: 10.1051/apido:2004044

[29] Al-Waili N, Salom K, Al-Ghamdi A, Ansari MJ. Antibiotic, Pesticide, and Microbial Contaminants of Honey: Human Health Hazards. The ScientificWorld Journal. 2012;9: 930849. DOI:10.1100/2012/930849.

[30] Mutinelli F, Costa C, Lodesani M, Baggio A, Medrzycki P, Formato G, Porrini C. Honey bee colony losses in Italy. Journal of Apicultural Research 2010;49(1):119-120. DOI: 10.3896/IBRA.1.49.1.24

Chapter 4

Chocolate: Health, Processing, and Food Safety

Ahmed Albandary, Fatemah Albandary and Amit K. Jaiswal

Abstract

Chocolate is a popular food product internationally, and it is consumed daily. Consuming chocolate has been linked to many human health benefits such as lower cholesterol levels, but there are some negative impacts such as weight gain because of its sugar content. Moreover, food safety issues related to chocolate have existed, and it can be contaminated by any biological, chemical, or physical hazards, which lead to many health issues. Regarding that, this chapter will discuss the benefits and negative impacts of consuming chocolate and provide the process of manufacturing the product.

Keywords: chocolate, health benefits, health risks, processing, food safety

1. Introduction

The cocoa bean is the seed of the cacao tree, it is a tropical plant indigenous to the equatorial regions of the Americas, but currently, it is cultivated in many countries with a warm, tropical climate. Cocoa is the main raw material to produce chocolate, which can value more than $100 billion [1]. Chocolate history started with the Maya, who were the first people who cultivated the cocoa plant in South America. During that period, chocolate was a cocoa drink prepared with hot water, and it can be flavored usually with cinnamon and pepper [2].

The Republic of Côte d'Ivoire, a country on the southern coast of West Africa, is the biggest producer of cocoa beans with more than 2 million tons annually. The national economy of Ivoirians depends on their export revenue from cocoa. Its population is more than 26 million, almost 6 million work in cocoa beans production, and almost 2,034,000 tons were produced in 2021 [1]. Many countries such as Ghana (883,652 tons), Indonesia (659,776 tons), Nigeria (328,263 tons), Cameron (295,028 tons), Brazil (235,809 tons), Ecuador (205,955 tons), Peru (121,825 tons), Dominican Republic (86,599 tons), and Colombia (56,808 tons) [3] that are considered top producers of cocoa come after Cote d'Ivoire.

This chapter is focused on the health benefits of chocolate, processing, type of chocolate, and finally, the safety aspect of chocolate production. Furthermore, the negative impact of chocolate on human health has been discussed.

2. Health benefits

There are numerous health benefits associated with the consumption of chocolate, which are due to its nutritional contents [4, 5]. A bar of chocolate with a higher number of nutrients has high health benefits, for example, dark chocolate [5], which contains soluble fiber and is rich in minerals. In more detail, a 100-gram bar of dark chocolate with 70–85% cocoa consists of 11 grams of fiber, 67% of the daily value (DV) for iron, 58% of the DV for magnesium, 89% of the DV for copper, 98% of the DV for manganese. So, dark chocolate is the best type with regard to the number of nutrients [6].

Chocolate possesses antioxidant activity, which is very beneficial for human health [5] and can be attributed to the presence of polyphenols, flavanols, and catechins, among others [6]. According to Nora et al. [7], antioxidants play an important role in the enhancement of immunity and for preventing cardiovascular diseases and metabolic diseases including obesity, diabetes mellitus, and some types of cancer. According to Harvard T.H. Chan School of Public Health [8], numerous studies showed a link between consuming chocolate and health benefits. Studies showed that consumption of 6 grams of chocolate daily leads to reducing the risk of heart disease and mortality, which could be due to reduced blood pressure and inflammation.

It is considered that chocolate is one of the 30 best anti-inflammatory foods. It has been observed that inflammation could lead to many health problems, and chocolate works as an anti-inflammatory effect in the human system, subsequently protecting humans against diseases. A study carried out by Creveling [9] found that cocoa polyphenols help to regulate the bacterial composition of the intestine. The author observed that cocoa polyphenols increase the number of good gut bacteria and that trigger this anti-inflammatory response [9].

Consuming chocolate could help in maintaining cardiovascular health. According to a study, carried out on 49 adults (32 women, 17 men) to assess the effectiveness of consuming dark chocolate on cholesterol levels, the results indicated that consuming chocolate that contains plant sterols and cocoa flavanols as a low-fat diet could contribute to cardiovascular health by reducing cholesterol and improving blood pressure [10].

Another study proved that the flavanols in dark chocolate can boost the endothelium, the lining of arteries, which lead to producing nitric oxide that works to send signals to the arteries to relax, which lowers the resistance to blood flow, and as a result of this, it reduces blood pressure [6]. Moreover, a study from Walden University's School of Nursing has been done on participants who eat dark chocolate, and it has been found that consumption of dark chocolate reduces blood pressure [11].

Consumption of dark chocolate improves the function of the brain. A study that was carried out on healthy volunteers confirmed that consuming high flavanol cocoa for 5 days contributes to improved blood flow to the brain. Furthermore, it also helps in improving cognitive function in older adults with mild cognitive impairment [6]. Nordqvist [5] referred that researchers from Harvard Medical School indicated that drinking two cups of hot chocolate a day for older people could help in keeping the brain healthy and reduce memory decline. They have found that hot chocolate helped in improving blood flow to parts of the brain where it was needed. A similar study was published in 2014 showed that a cocoa extract that is called lavado could help in reducing or preventing nerve pathways found in patients with Alzheimer's disease and could contribute to slow symptoms such as cognitive decline [5].

Consumption of chocolate also helps to protect human skin against skin disease because of cocoa flavanols [6, 11, 12]. Also, chocolate contains theobromine, which helps pregnant women against complications of pregnancy, which is known as preeclampsia. According to a study that was done on pregnant women, as they were given a higher amount of chocolate, they had a 40% less chance of getting preeclampsia [13]. Studies also showed that consuming chocolate can help in protection against cancer as Brook [14] indicated in a study, which was published in the Journal of the American Society of Hypertension, that people who consume many flavonoids or antioxidant-rich chocolate have lower chances of developing cancer, compared with those who do not.

3. Side effects associated with chocolate consumption

As discussed in the last section, there are numerous health benefits associated with the consumption of chocolate; however, there are also some negative effects on human health such as childhood hyperactivity, migraine, and headaches. According to a study conducted in the Bogalusa Heart, to identify the presence of caffeine in children's snacks, the results indicated the presence of a large amount of caffeine in children, which could be due to eating foods such as chocolate [15]. Bruso [16] referred to that according to the University of Maryland Medical Center, approximately 8% of children in the US younger than 18 have hyperactivity at some point. These cases are often treated by behavioral therapy and medication, but some parents resort to changing children's diets by avoiding chocolate in their children's meals. The Medical Center reported that this change is an effective method but only in a small percentage of children [16].

According to Nowaczewska et al. [17], there are 25 studies have indicated the prevalence of chocolate as a migraine trigger. Two studies of them indicated that chocolate is not a migraine trigger because no participant has reported that chocolate is a trigger factor. Also, other studies considered chocolate a migraine trigger in a small percentage of participants (ranging from 1.3 to 33) Nowaczewska et al. [17]. According to Kelishadi [15], there are some authors who say migraine and tension-type headache occur due to the consumption of some type of foods: one of them is chocolate. Also, another study indicated that chocolate could provoke a migraine attack in some patients who think themselves sensitive to it. [15]. An increase in migraines in some people upon eating chocolate could be due to cocoa's tyramine, histamine, and phenylalanine content [5]. It has been also reported that consuming more chocolate could lead to bone health issues. Nordqvist [5] mentioned in a study, which was published in The American Journal of Clinical Nutrition, low bone density in older women who consume chocolate every day. Also, it has been reported that some chocolate includes a high amount of heavy metals such as cadmium and lead, which can affect the kidneys and bones [5]. A study found that almost all samples (43 chocolate products) contain more than 0.3 mcg of cadmium per serving, which means more than the recommended level identified by the World Health Organization [5].

4. Production of chocolate

The cacao tree is considered a sensitive plant, and it requires protection from wind and needs shade in its early years [18]. The cocoa tree needs a semi-shaded environment (50% light and 50% shade). In the cocoa plantations, there are many tall trees beside cocoa trees to provide the required shade and shall not prevent the light [19]. The cocoa plant also needs high rainfall and temperatures to grow. Therefore, it can only grow in a narrow band of countries between 20 degrees north and south of the equator [20]. The suitable temperature for growing is between 25 and 27°C (between 77 and 81°F). The tree cannot resist the high dry or wet. The perfect rainfall should be between 1250 and 2500 mm per year. The cocoa tree requires fertile, slightly acidic, and well-drained soil [21].

The wild cocoa tree can grow up to 12–15 m but to make harvesting easier, most farmers grow it to less than 5 m. Regarding blooms, it blooms and bears fruit all year round, which means that the cocoa tree has both flowers and fruits at the same time [21]. Moreover, the most producing period is from October to February and from May to August [22].

The products could be different in many properties such as shape, texture, and size. The length can be started from 15 cm to over 35 cm. One ripe fruit contains 20–75 cocoa beans, the length of each one can be from 1 cm to 3 cm, embedded in a white pulp. The nutritious cocoa bean is considered very high beneficial as it contains fat (50%) and carbohydrates (25%), as well as proteins, theobromine, niacin, vitamins (including A, B1, B2, and B6), and minerals (calcium, iron, potassium, magnesium, sodium, and phosphorus) [21].

4.1 Varieties of cocoa

There are three types of chocolate, which are Criollo, Forastero, and Trinitario. These types are dominant varieties [23]. It was named officially in 1753 by the Swedish scientist Carl von Linné [24]. According to the chocolate society [25], the most productive and common type is Forastero beans, which are because of their high yielding in Brazil and W Africa. On the other hand, Criollo and Trinitario beans are better in quality, but they are lower yielding, which made them very expensive.

4.1.1 Trinitario

This type of cocoa exists on the Island of Trinidad after a hurricane nearly destroyed the local Criollo crops in 1727. As a result, farms are replanted with the Forastero type, which was brought from Venezuela, and cross-fertilized with the native Criollo type [25]. Therefore, a new type appears, which is known now as Trinitario. This type merged the best two types, which are Forastero (which is featured by the hardiness and is very high in yielding) and the Criollo type (which is featured by great taste) [21]. As a result, the Trinitario variety is classified as fine flavor cocoa, but it is less intense than the Criollo type [26]. The properties of this type are usually varied because of the parents as they have different characters, and it can be grown where Criollo cocoa was once grown including in Mexico, the Caribbean islands, Colombia, Venezuela, and parts of Southeast Asia [21]. The percentage of the world's production is uneven regarding the Walter Matter [26] approximately only 5% of the world's production, and according to Bar & Cocoa, proved that the production is less than 10% globally. However, according to the chocolate society [25], this type represents about 12% of the world's cocoa production, and in 2020, the Ministry of Foreign Affairs confirmed that the production represents approximately 10–15%.

4.1.2 Criollo

This type was cultivated by the Maya more than 2500 years ago. It is considered a high-quality variety compared with others. Also, it is very complex with regard to cultivation and handling; consequently, it represents less than 5% globally, and it is the rarest type [27]. It is cultivated in Central America, northern South America, the Caribbean, and Sri Lanka. It has a delicate and sweet flavor, so it is often mixed with different types because of its rarity and high cost [28]. According to its characteristics, the pods are red or green before ripeness with less than ½ in length, and it has a very accentuated tip at the lower end, marked with 10 deep furrows. Its surface is smooth and round-shaped. Moreover, it is plump with fresh white cotyledons, and it is very easy to ferment [27].

4.1.3 Forastero

This is the most common type of cocoa, its native most likely to the Amazon basin. These days, it is grown mostly in Africa, Ecuador, and Brazil. It is represented almost 80% of the world's cocoa production, because of its properties, which include hardness, and resistance to diseases. There are many subspecies of Forastero such as Cundeamor and Calabacillo [21]. This type was discovered by the Spanish in the heart of the Amazon [27]. It grows in West Africa and Brazil [26].

4.2 Diseases of cocoa fruit

4.2.1 Black pod

It is mostly cultivated cocoa in small labor-intensive farms for less than 2 hectares (5 acres) to protect the product against disease and pests. Even though, still, there are many losses of production globally, which approximately range from 30 to 100%. The most popular diseases of the cacao tree are pod rots. It is called a black pod as a result of fungus (Phytophthora), which spreads very quickly on the pods under rainy and humid climate conditions, with a lack of sunshine, and temperature under 21°C (70 °F). Therefore, control is required by using copper-containing fungicides, and the infected pods should be removed to reduce spread.

4.2.2 Witches' broom

Witches' broom, which is caused by *Moniliophthora perniciosa*, has a very dangerous effect on products [29]. It is one of the most risky diseases throughout South America. High temperature and high humidity are good environments for this disease. Therefore, it must be following good sanitation and remove any infected materials, which are difficult because it could not show any visible symptoms. Also, fungicides must be used to protect the cocoa [30].

4.2.3 Frosty pod

This disease is caused by *Moniliophthora roreri* fungus, and it is considered a highly infectious disease that impacts the production of cocoa. It is spread rapidly by water movement, wind, or the movement of pods [30].

4.2.4 Swollen shoot

This one is only found in West Africa, which is a huge problem in Togo, Ghana, Cote d'Ivoire, and Nigeria. It is transmitted by mealybugs. This natively results from trees that grew in the rain forests of West Africa, so it is not native to the cocoa plants [30].

Table 1 includes the main estimated losses of cocoa production annually because of diseases [31].

4.3 Pest

There are over 1500 different insects that feed on cocoa; however, there is only 2% of economic importance. Mirid bugs such as Helopeltis are the most common and important insects in attacking cocoa, and the main pest in Malaysia and Indonesia is the cocoa pod borer. The most popular insect pests are including broad mite, flower-eating caterpillars, Helopeltis, and yellow peach moth [31].

Disease	Region	Estimated world production loss (tonnes)
Black pod (phytophthora fungus)	Africa, Brazil, Asia	450,000
Witches broom (fungus)	Latin America	250,000
Frosty pod rot (fungus)	Latin America	30,000
Swollen shoot virus	Africa	50,000
Vascular streak dieback (virus)	Africa	30,000

Source: https://www.daf.qld.gov.au/business-priorities/agriculture/plants/fruit-vegetable/fruit-vegetable-crops/cocoa/pests-and-diseases-of-cocoa.

Table 1.
Estimated losses of cocoa production annually because of diseases.

5. Processing of cocoa

5.1 Harvesting cocoa

The coco plant life on average is 25 years, and the product needs range from 150 and 165 days for complete maturation. There is no specific season for ripening as it can be harvested all year round [19]. Beans grow in pods that sprout out of the trunk and branches of cocoa trees. The size of the pods is almost the same as the football size. The pods start green and then convert to orange, which means they are ripe and ready for harvesting [32].

When the pods are ripe, harvesting should be done manually, as using machines will lead to damage to the tree, the clusters of flowers and pods that grow from the trunk [32], as the tree is weak, which makes the picking a hard job for workers [18]. For picking, workers use short, hooked blades mounted on long poles to reach the highest products [32]. After picking cocoa from trees, it is collected in baskets and transported to the braking operation [32]. In this step, it is split, and the cocoa beans are removed. Pods can contain upward of 50 cocoa beans each [18]. The pods are opened, and the beans are removed within a week to 10 days after harvesting to avoid germinating [33].

5.2 Fermentation of cocoa

Fermentation of cocoa after harvesting is a critical process [34]. This process leads to getting the aromas to the cocoa and knowing the difference between many types of cocoa. However, for some traditional dishes, unfermented beans are used in parts of Mexico and Central America. The fermentation method could take 2–8 days; it depends on the type of cocoa [18]. It is fermenting the pulp of Forastero type for 5–7 days, and the pulp of Criollo type for 1–3 days [29].

During fermentation, the cocoa beans are placed in large, shallow, heated trays or covered with large banana leaves. If the weather is good, it is heated by the sun with the importance of noticing that it is necessary to move them up frequently to allow the beans to come out for fermentation equally [32]. In this method, the juicy sweating of the pulp is drained and the germ in the seed is killed by the heat, and the flavor will be improved. As well as the color will be changed to brown. This step is implemented in sun-dried or kiln-dried to reduce the moisture content to 6–7 % [29]. Throughout the fermentation of cocoa beans, the top layer of the beans is covered with banana leaves, because the bottom part of the banana leaf consists of natural yeast and microorganisms, which help to strengthen the natural fermentation [18].

5.3 Drying

After fermentation, a drying process must be implemented, which is an important step to enhance the cacao flavor [35]. The drying is done by the sun, and it should turn the beans as well in this step to be equally dried. It can dry on wooden floors, which can be covered by a sliding roof during rainy weather. Electric dryers are used on large farms [18]. The drying stage period could take between 2 and 10 days in the nature processing [36]. During drying, the color of cocoa beans is changed from reddish-brown to dark brown [18]. As a result of this process, the humidity of the cocoa is reduced from 60% to 7% [35].

In this method, the polyphenol content of cocoa beans is reduced, many numbers have been reported in a review study, which was done by [37], and depending on many studies include a reduction from 77 to 44%, a 72% reduction, a 30% reduction, and a 26% reduction. In addition, sun drying reduces the phenolic content and antioxidant activity of cocoa beans.

However, according to a study that was done by [36] to compare the methods of drying cocoa beans to evaluate antioxidant activity, to determine phenolic compounds and methylxanthine content, and to determine the presence of ochratoxin A, four methods are used:

Dryer with stainless steel platform and plastic roof with UV protection (DP):

Artificial dryer, using the wooden platform with an artificial heat source, through forced and heated air circulation with electric resistance and temperature controlled by a thermostat at 60°C (AD);

Traditional dryer in the barge with wooden platform and drying by direct sunlight (TD);

Mixed dryer with stainless steel platform and mobile plastic roof with UV protection for drying coverage and exposure to sunlight (MD).

It has been found that the best method is the traditional drying method, in which drying is by the sun directly to maintain the phenolic compound content in the cocoa beans, as the phenolic compound content is reduced in different methods, particularly in the artificially oven-dried method. Also, the traditional method is the best in retention of cocoa antioxidant activity and the methylxanthine content in dry seeds, which confirm that these chemical compounds are impacted by changing temperature. The reason for the traditional method being the best could be because it uses less than milder temperatures.

5.4 Manufacturing process of chocolate

5.4.1 Cleaning

The first process that must be done in the industry is cleaning the cocoa beans probably by using sieves and brushes [19]. There are many cleaning steps to remove any contamination, for example, twigs, stones, and dust [29]. In general, the manufacturing operations are different slightly because of the various types of cocoa trees; however, most industries use similar machines to convert cocoa beans to cocoa butter and chocolate [32].

5.4.2 Roasting

The roaster step is very important in processing chocolate, and that is done by drying beans at a temperature of 100°C, then roasting beans at a temperature range between 100°C and 160°C. It could be different depending on the type. The balance in roasting is very necessary as increasing it will lead to the bitterness of the beans and less than required will not improve the aromas [19]. A study [38] has proved that roasting at high temperature influences cocoa beans and has no influence on roasting time. Moreover, roasting at 160°C leads to undesirable burnt odor and flavor and a low acceptability score by consumers. However, roasting at temperature range from 90 to 110°C was acceptable by consumers in appearance, aroma, flavor, texture, and overall quality attributes.

This process develops the aromas and contributes to reducing acidity and astringency, reducing moisture content, deepening color, and facilitating shell removal [29]. Also, roasting is killing any organisms that exist on the bean, which appear in the fermentation process [18]. In this process, it has been found to decrease the flavanols and phenolic content of cocoa beans [37].

5.4.3 Grinding

The roasting method makes the shells of the cocoa brittle, so it is winnowed to remove the shells of the bean and only cocoa nibs pass through a series of sieves, and this method is called winnowing [29]. In the grinding process, the nibs are ground in a granite stone mill, which crushes the grain and releases the fat or cocoa butter [29]. Cocoa butter is the main ingredient of chocolate [19]. Chocolate consists of both cocoa solids and cocoa butter in almost the same ratio [18].

5.4.4 Coaching

This process is completed by conche machines. Moreover, this process is developing flavor, aerating, and emulsifying. The required time is 4–72 hours (depending on the desired results, and the machine type plays a major as well). The temperature is between 55 and 88°C (130 and 190 °F) and is controlled regarding the desired flavor and uniformity [29]. In this process, chocolate could lose almost 80% of volatile substances [19].

5.5 Types of chocolate

There are many types of chocolate that are different in ingredients and characteristics.

1. Dark chocolate, which contains the highest percentage of cocoa bean solids (up to 80% of the total weight) and cocoa butter compared with other varieties. This type is featured by the strong aroma of cocoa. Its quality depends on the percentage of cocoa. Most of the human benefits of health have been linked to consuming this type.

2. Another type is called Gianduja chocolate, which contains hazelnuts, cocoa, and sugar. Its color is brown.

3. Milk chocolate consists of cocoa butter, sugar, milk powder, lecithin, and cocoa of 20–25%. Its properties are including a bright appearance, an intense, persistent aroma, and a sweet taste with a slightly bitter accent of cocoa [2]. According to FDA Standards of chocolate identity, milk chocolate must include at least 10% chocolate liquor and 12% milk solids. Also, cocoa butter and milk are the only fat allowed in this type [39].

4. The last type is white chocolate, which contains cocoa butter, milk, and sugar with no cocoa solids. It is a sweet, pleasant taste [2]. To meet the FDA standards, this type must have at least 20% cocoa butter and 14% milk solids [39].

6. Food safety of chocolate

Food contamination is referring to food that is contaminated microbiologically, chemically, or physically. That can be done at any stage of food processing such as during storage or transportation [40]. Also, food allergies can be considered a contamination factor [41].

Food safety issues are considered a huge challenge in the health sector compared to those of malaria or tuberculosis [42]. As the person who consumes contaminated food gets food poisoning and starts the poisoning symptoms in hours, and it could need to visit a hospital, particularly the high-risk group, which includes older adults, pregnant women, infants, and young children, and people with chronic disease [43]. Food poisoning includes many common symptoms, which are upset stomach, stomach cramps, nausea, vomiting, diarrhea, and fever [44]. There are many foods that are linked to food poisoning, which are poultry, raw fruits and vegetables, fish and shellfish, rice, deli meats, unpasteurized dairy, and eggs [45]. Many cases of outbreaks of food poisoning in the United States have been linked to fruits and vegetables in 2018 [46].

6.1 Microbial contamination of food

Microbial contamination is the most common type of food contamination. It can be done if the food is contaminated by microorganisms such as bacteria, viruses, mold, fungi, and toxins. Many reasons lead to microbial contamination that including undercooking chicken, storing and preparing raw foods near to ready-to-eat food, and this leads to cross-contamination [41]. Also, the vectors of biological contamination include food handlers and that can be caused by aerosol droplets from coughing near the production line. Moreover, vectors include packaging materials, equipment, and tools used; an example of this is using the same cutting board and knife in raw food and ready-to-eat food. Contaminated water is also an important factor to cause microbial contamination. Furthermore, pests such as insects and rodents are factors of contamination [40]. Therefore, food hygiene practices must be implemented to prevent contamination that including personal hygiene, separating raw and ready-to-eat food at all stages of processing, washing raw fruits and vegetables, and pest control on the premise [41].

For a long period, it was referred to low-moisture foods, such as chocolate, as a safe product against microbial contamination because of their properties, as the water activity (aw) is below 0.6, which is not an active area for microbial growth. However, the early 1970s were linked to the first outbreaks of Salmonella to low-moisture products such as chocolate, oat cereals, peanut butter, and infant formula. Salmonella was the main pathogen of concern for all of those products [47]. Also, another study that was to identify the emerging hazards related to chocolate products between 2013 and 2018 has concluded that microbiological hazard is 16,49%, and it is the second hazard following chemicals in cocoa from Africa and South America as well [48]. According to the most recent data, it has been shown that a Salmonella outbreak in a brand of chocolate wafers from Poland between December 2020 and early April 2021affected 32 people as announced [49].

Moreover, many reasons cause biological contamination linked to chocolate products, for example, pathogenic microorganisms can be found because of using contaminated ingredients such as milk and sugar. Also, pathogens can be presented because of damage or soiled packaging material because of mishandling at the supplier level. Furthermore, it can be found in returned chocolate products because of mishandling in any area such as retail or during transportation [50].

6.2 Chemical contamination of food

Chemical contamination is any food contaminated by a chemical agent. For example, food is contaminated by cleaning agents and that can happen when using chemical cleaning during processing or when using chemical cleaning on tools and not cleaning well. Furthermore, there is natural chemical contamination such as toxins in some fish, such as methylmercury [51]. Also, contamination can happen in agriculture production when they spray fertilizers and pesticides close to food when it is growing. Therefore, it must be ensured that chemical agents are away from the food area. Always follow the manufacturers' instructions before using chemicals. Food must be covered when cleaning and dealt only with approved chemical suppliers [41].

There are many potential chemical contaminations related to chocolate that can happen at any stage of the food chain, which includes the existence of environmental contaminants such as pesticides above the allowance level, as well as presence of food additives in dark chocolate or heavy metals or addition of food additives, which are not permitted for use by government regulations [50].

This type of contamination, either industrial contaminants or pesticides, is most commonly related to chocolate products in Africa, South America, and Asia [48]. Another study proved that some cocoa products exceed the European Union and Chinese Maximum Contaminant Level regarding arsenic, cadmium, lead, and mercury, which could affect human health [52].

6.3 Physical contamination of food

Physical contamination of food is any food that contains a foreign object at any stage of the production process. That foreign substance causes health issues for consumers, such as broken teeth or choking [41]. There are many food products recalled every year because of physical contamination, which causes consumers health issues and costs less. There are many real examples from the Food and Drug Administration (FDA) 2016 product recall list such as "small metal shavings in apple coffee cakes, metal fragments in sugar used in Asian sauce, plastic mesh screen fragments in flour, and pieces of rubber in baby food" [53].

An example of physical contamination in almond processing is that, when the crop is ready for harvesting, the tree is shacked by a tractor machine, which leads to falling of almonds on the ground, which are left for many days to dry. Then, another machine sweeps them into rows so a harvester can pick them up with a series of belts. The almonds are cleaned by the machine and dumped into a bucket in the back of the truck. Then the truck is transferred to the industry to remove the shells and sticks and clean the product. The industry machines can be damaged and worn out, which could lead sometimes to the transfer of small pieces of that machinery into a package. This could include bolts and washers. Therefore, using technology in processing plants is an important role such as using X-rays and sensors and food metal detectors to identify any foreign substance to protect humans' health [53].

6.4 Allergenic contamination of food

Food allergies are an individual adverse reaction to some food. Food allergens are proteins that can be found in food in huge amounts [54]. Even a small amount of food is enough to impact a person who has a food allergy [41]. According to the Food Safety Authority of Ireland [55], there are 14 named allergens (**Table 2**).

Allergic food contamination can happen when food from the list of food allergies enters or contacts another food. An example of this is using the same container to store pasta that is already selected to store peanuts. Therefore, it must be ensured to

Food allergen products	
1	Cereals containing gluten such as wheat
2	Crustaceans and products thereof
3	Eggs and products thereof
4	Fish and products thereof
5	Peanuts and products thereof
6	Soybeans and products thereof
7	Milk and products thereof
8	Nuts, which are almonds, hazelnuts, walnuts, cashews, pecan nuts, Brazil nuts, pistachio nuts, macadamia or Queensland nuts, and products thereof
9	Celery and products thereof
10	Mustard and products thereof
11	Sesame seeds and products thereof
12	Sulfur dioxide and sulfites at concentrations of more than 10 mg/kg
13	Lupin and products thereof
14	Molluscs and products thereof

Table 2.
Food allergies products.

deal only with approved suppliers who take critical control of allergenic contamination. Always separate tools, equipment, and clothes that are used for allergenic foods from other foods. Separate allergen in storage from other foods. Cleaning must be implemented properly in the area after the use of one of the 14 allergen products.

Many food safety concerns can be avoided when implementing an effective prerequisite program that includes cleaning and sanitation, staff training, maintenance, chemical control, waste management, and storage and transportation [56]. Hygiene must be implemented in the chocolate industry to provide safe products to protect consumers [57].

It should provide maintenance of food equipment and infrastructure; this can be done by regularly auditing all equipment used in the chocolate factory and fixing any problem, which could post hygiene issues and food safety. Also, hygiene must be maintained during maintenance and the equipment cleaned after that and sanitized before using it again, and the product carefully protected from any foreign body. For the infrastructure part, it should follow a regular check to protect the production area from any issues such as soil and dust to keep the environment safe and hygiene. Any repair must be done outside of production periods, but if that is not possible, a separation wall must be set to protect the production area [57].

Also, it must follow cleaning and disinfecting daily in the industry and removing any undesirable substances such as residues and foreign bodies, which could affect the hygiene of products or equipment. Also, it must separate chemical agents apart from food products to prevent cross-contamination, and all storage should be cleaned regularly [57].

In general, food industries, particularly the chocolate industry attracts insects and rodents through the odors as this place is seen as a great environment for insects because of the presence of food and water and shelter. Therefore, pest control must be implemented effectively, and auditing inspection should be followed regularly and correct any failure related to that. Moreover, it must protect the chocolate production from any contamination, so it should set rules for staff working and visitors such as regularly washing their hands and wearing gloves, mouth masks, hairnets, and clean disinfected clothing [57]. Regarding peanuts and other nuts, they should be segregated as much as possible. It should clean any machine correctly during changing from nut products to other products and must refer to the label that this industry is processing nuts. A further step that should be implemented is using a metal detector to ensure that the product is free from any foreign body such as plastic or other [58].

Furthermore, applying a Hazard Analytical Critical Control Point is required to prevent any type of contamination as this system is to reduce or prevent any risks early before it happens [56]. A Critical Control Point (CCP) is referred to as a step that can be applied, and it is a necessity to prevent or reduce food safety hazards. Most steps include a heating, cooking, or cooling stage. Examples of critical control points in chocolate production include roasting cocoa at temperatures between 105°C and 120°C and at specific times to eliminate pathogens. The further step is metal detection.

Also, there are many control points in chocolate production, for example:

- Receiving, which is required visual inspection, and use of chemical-free materials
- Splitting; use sanitized equipment and personal hygiene.
- The fermentation that is required visual inspection, and regular maintenance of fermentation tanks.
- Drying, which needs visual inspection, observation of time, and temperature.
- Roasting is very important to follow the correct time and temperature to kill pathogens.
- Grinding required also visual inspection and sanitization of equipment.
- Coaching that includes visual inspection and proper maintenance of equipment.
- Tempering—visual inspection, sanitized equipment
- Molding—ensure molds are clean.
- Packing—metal detection, food-grade inks, appropriate packaging materials, correct labeling [59].

7. Conclusions

In conclusion, cocoa is an important product globally as it provides many health benefits for people such as reducing the risk of heart disease, reducing blood pressure, and working as an anti-inflammatory substance. However, consuming too much could lead to many negative impacts on human health such as diabetes or overactivation for children because of the presence of caffeine and weak bone for elderly people, subsequently should avoid eating a lot of chocolate. Also, more research is needed to establish more health benefits. Chocolate is considered a low-risk product, but there are many contaminants linked to it, which could cause many health issues; therefore, good hygiene practices and the HACCP system should be followed to provide safe products.

Conflict of interest

The authors declare no conflict of interest.

Author details

Ahmed Albandary[1]*, Fatemah Albandary[2] and Amit K. Jaiswal[3]

1 Food Services Department, Quality Assurance and Food Safety Unit, King Fahd University of Petroleum and Minerals, Saudi Arabia

2 Independent Researcher, Saudi Arabia

3 School of Food Science and Environmental Health, College of Sciences and Health, Technological University, Dublin, Ireland

*Address all correspondence to: a.albandary1990@gmail.com

References

[1] Shvili J. The Top Cocoa-Producing Countries. 2020. Available from: https://www.worldatlas.com/articles/top-10-cocoa-producing-countries.html [Accessed: September 20, 2021]

[2] Montagna et al. Chocolate, "Food of the Gods": History, Science, and Human Health. 2019. Available from: https://www.mdpi.com/1660-4601/16/24/4960 [Accessed: September 20, 2021]

[3] World Population Review. Cocoa Producing Countries. 2021. Available from: https://worldpopulationreview.com/country-rankings/cocoa-producing-countries [Accessed: September 20, 2021]

[4] Cleveland Clinic. Heart Healthy Benefits of Chocolate. 2021. Available from: https://my.clevelandclinic.org/health/articles/16774-heart-healthy-benefits-of-chocolate [Accessed: August 5, 2021]

[5] Nordqvist J. Health benefits and risks of chocolate. 2018. Available from: https://www.medicalnewstoday.com/articles/270272 [Accessed: August 4, 2021]

[6] Gunnars K. 7 Proven Health Benefits of Dark Chocolate. 2021. Available from: https://www.healthline.com/nutrition/7-health-benefits-dark-chocolate [Accessed: August 4, 2021]

[7] Nora et al. Fruits and Vegetables in the Management of Underlying Conditions for COVID-19 High-Risk Groups. 2021. Available from: https://www.mdpi.com/2304-8158/10/2/389 [Accessed: August 20, 2021]

[8] Harvard T.H. Chan School of Public Health. Dark Chocolate. 2021. Available from: https://www.hsph.harvard.edu/nutritionsource/food-features/dark-chocolate/ [Accessed: August 7, 2021]

[9] Creveling M. 14 Health Benefits of Eating Chocolate. 2021. Available from: https://www.eatthis.com/chocolate-health-benefits/ [Accessed: August 7, 2021]

[10] Allen et al. Daily Consumption of a Dark Chocolate Containing Flavanols and Added Sterol Esters Affects Cardiovascular Risk Factors in a Normotensive Population with Elevated Cholesterol. 2008. Available from: https://academic.oup.com/jn/article/138/4/725/4665024 [Accessed: August 5, 2021]

[11] Mertes A. Is Chocolate Good for You? The Pros & Cons of Eating Chocolate. 2021. Available from: https://www.qualitylogoproducts.com/blog/health-benefits-of-chocolate/ [Accessed: August 4, 2021]

[12] Realbuzz Team. 7 Health Benefits Of Chocolate. 2021. Available from: https://www.realbuzz.com/articles-interests/nutrition/article/7-health-benefits-of-chocolate/ [Accessed: August 15, 2021]

[13] Locke R. 20 Health Benefits Of Chocolate. 2021. Available from: https://www.lifehack.org/articles/lifestyle/20-health-benefits-chocolate.html [Accessed: August 4, 2021]

[14] Brooks A. 8 Healthy Reasons to Eat Dark Chocolate. 2019. Available from: https://www.everydayhealth.com/diet-and-nutrition-pictures/delicious-reasons-to-eat-dark-chocolate.aspx [Accessed: August 8, 2021]

[15] Kelishadi R. Cacao to cocoa to chocolate: Healthy food?. 2005. Available from: https://sacredchocolate.com/docs/sacredpdf/cacao-cocoa-chocolate-arya-journal.pdf [Accessed: August 8, 2021]

[16] Bruso J. Chocolate's Influence on Hyperactivity in Children. 2011.

Available from: https://healthfully.com/159485-chocolates-influence-on-the-hyperactivity-of-a-child.html [Accessed: August 10, 2021]

[17] Nowaczewska et al. To Eat or Not to eat: A Review of the Relationship between Chocolate and Migraines. 2020. Available from: https://www.ncbi.nlm.nih.gov/pmc/articles/PMC7146545/ [Accessed: August 10, 2021]

[18] Science of Cooking. How is Chocolate Made?. 2021. Available from: https://www.scienceofcooking.com/chocolate/how-is-chocolate-made.htm [Accessed: July 7, 2021]

[19] Cioccolat Italiani. Production of chocolate, from the bean to packaging. 2021. Available from: https://www.cioccolatitaliani.it/en/blog/production-of-chocolate-from-the-bean-to-packaging.html [Accessed: July 7, 2021]

[20] Cocoa Life. COCOA GROWING. 2021. Available from: https://www.cocoalife.org/in-the-cocoa-origins [Accessed: July 7, 2021]

[21] Barry Callebaut. Theobroma cacao, the food of the gods. 2021. Available from: https://www.barry-callebaut.com/en/group/media/press-kit/theobroma-cacao-food-gods [Accessed: July 5, 2021]

[22] Singh R. Cacao. 2021. Available from: https://www.britannica.com/topic/cocoa-food [Accessed: July 10, 2021]

[23] Espinet R. The story of Trinitarios. 2019. Available from: https://newsday.co.tt/2019/06/26/the-story-of-trinitarios/ [Accessed: July 10, 2021]

[24] Willie's Cacao Ltd. The different cocoa varieties. 2021. Available from: https://www.williescacao.com/world-cacao/different-cacao-varieties/ [Accessed: July 13, 2021]

[25] The Chocolate Society. The different varieties of cocoa beans: Criollo, Trinitario, & Forastero. 2021. Available from: https://www.chocolate.co.uk/blogs/news/the-different-varieties-of-cocoa-beans-criollo-trinitario-and-forastero [Accessed: July 10, 2021]

[26] Walter Matter (2021) Varieties of Cocoa. Available from: https://www.wama.ch/en/varieties-of-cocoa.html [Accessed: August 15, 2021]

[27] Chococlic. Criollo Cocoa. 2021. Available from: https://en.chococlic.com/?s=Criollo [Accessed: July 13, 2021]

[28] Ministry of Foreign Affairs. The European market potential for speciality cocoa. 2020. Available from: https://www.cbi.eu/market-information/cocoa-cocoa-products/speciality-cocoa/market-potential [Accessed: July 13, 2021]

[29] Cook L. Cacao. 2021. Available from: https://www.britannica.com/plant/cacao [Accessed: July 10, 2021]

[30] Plant Village. Cocoa. 2021. Available from: https://plantvillage.psu.edu/topics/cocoa-cacao/infos [Accessed: July 14, 2021]

[31] Queensland Government. Pests and diseases of cocoa. 2013. Available from: https://www.daf.qld.gov.au/business-priorities/agriculture/plants/fruit-vegetable/fruit-vegetable-crops/cocoa/pests-and-diseases-of-cocoa [Accessed: July 14, 2021]

[32] Tyler et al. The World Atlas of Chocolate. 2003. Available from: https://www.sfu.ca/geog351fall03/groups-webpages/gp8/intro/intro.html [Accessed: July 3, 2021]

[33] Cocoa and Heart. How are cocoa beans harvested. 2021. Available from: https://www.cocoaandheart.co.uk/how-are-cocoa-beans-harvested/ [Accessed: July 18, 2021]

[34] Vuyst L, Weck S. The cocoa bean fermentation process: from ecosystem

analysis to starter culture development. 2016. Available from: https://sfamjournals.onlinelibrary.wiley.com/doi/pdf/10.1111/jam.13045 [Accessed: July 18, 2021]

[35] Guevara J. Cacao Harvesting & Processing: A Step By Step Explanation. 2018. Available from: https://perfectdailygrind.com/2018/02/step-step-explanation-cacao-harvesting-processing/ [Accessed: July 18, 2021]

[36] Deus et al. Influence of drying methods on cocoa (Theobroma cacao L.): Antioxidant activity and presence of ochratoxin A. Food Science and Technology. 2018;**38**:278-285

[37] Di Mattia et al. From cocoa to chocolate: The impact of processing on in vitro antioxidant activity and the effects of chocolate on antioxidant markers in vivo. Frontiers in Immunology. 2017;**8**:1207

[38] Rocha et al. Effect of the roasting temperature and time of cocoa beans on the sensory characteristics and acceptability of chocolate. 2017. Available from: https://www.scielo.br/j/cta/a/6NvBxynyYXyHKXpxsfxxqbG/?lang=en [Accessed: July 19, 2021]

[39] Vishnefske J. FDA Standard Definition of Chocolate. 2017. Available from: https://www.santabarbarachocolate.com/blog/fda-standard-definition-of-chocolate/ [Accessed: July 22, 2021]

[40] Abdolshahi A, Yancheshmeh B. Food Contamination. 2020. Available from: https://www.intechopen.com/books/mycotoxins-and-food-safety/food-contamination [Accessed: May 30, 2021]

[41] Collier E. What are the Four Types of Food Contamination?. 2019. Available from: https://www.highspeedtraining.co.uk/hub/four-types-contamination/ [Accessed: May 29, 2021]

[42] Hoffmann V. Addressing food safety questions along the value chain. 2019. Available from: https://www.ifpri.org/blog/addressing-food-safety-questions-along-value-chain [Accessed: May 30, 2021]

[43] Mayo Clinic Staff. Food poisoning. 2020. Available from: https://www.mayoclinic.org/diseases-conditions/food-poisoning/symptoms-causes/syc-20356230 [Accessed: May 29, 2021]

[44] CDC. Food Poisoning Symptoms. 2021. Available from: https://www.cdc.gov/foodsafety/symptoms.html [Accessed: May 29, 2021]

[45] Coyle D. Top 9 Foods Most Likely to Cause Food Poisoning. 2018. Available from: https://www.healthline.com/nutrition/foods-that-cause-food-poisoning [Accessed: July 1, 2021]

[46] Moreb et al. Fruits and Vegetables in the Management of Underlying Conditions for COVID-19 High-Risk Groups. 2021. Available from: https://www.mdpi.com/2304-8158/10/2/389 [Accessed: June 25, 2021]

[47] Winkler. Microbiological safety of chocolate confectionery products. 2009. Available from: https://www.newfoodmagazine.com/article/189/microbiological-safety-of-chocolate-confectionery-products/ [Accessed: July 6, 2021]

[48] Foodakai. Case study: Chocolate Product. 2019. Available from: https://agroknow.com/wp-content/uploads/2020/07/Foodakai_Case_Study_Chocolate_Industry_Report.pdf [Accessed: July 8, 2021]

[49] Whitworth J. Salmonella outbreak in Sweden linked to chocolate wafers. 2021. Available from: https://www.foodsafetynews.com/2021/04/salmonella-outbreak-in-sweden-linked-to-chocolate-wafers/ [Accessed: July 28, 2021]

[50] Food Ready. Food Safety / HACCP Plan—Chocolate. 2021. Available from: https://foodready.ai/food-safety-haccp-guidance/plan-guidance-chocolate/ [Accessed: July 25, 2021]

[51] Australian Institute of Food Safety. Food Safety and the Different Types of Food Contamination. 2021. Available from: https://www.foodsafety.com.au/blog/food-safety-and-the-different-types-of-food-contamination [Accessed: July 20, 2021]

[52] Anyimah-Ackah et al. Exposures and risks of arsenic, cadmium, lead, and mercury in cocoa beans and cocoa-based foods: a systematic review. 2019. Available from: https://academic.oup.com/fqs/article/3/1/1/5365839?searchresult=1#133623096 [Accessed: July 25, 2021]

[53] Schultz M. What are physical contaminants in food manufacturing?. 2018. Available from: https://www.thermofisher.com/blog/food/what-are-physical-contaminants-in-food-manufacturing/ [Accessed: May 29, 2021]

[54] Popping B, Bourdichon F. Allergen management in the food industry: it's all about cross-contact and contamination prevention. 2017. Available from: https://www.newfoodmagazine.com/article/73985/allergen-management/ [Accessed: May 29, 2021]

[55] Food Safety Authority of Ireland. List of 14 Allergens. 2014. Available from: https://www.fsai.ie/legislation/food_legislation/food_information/14_allergens.html [Accessed: May 31, 2021]

[56] Albandary and Moran. A Study of the Effectiveness of a Software HACCP System FoodCheck for the Management of Food Safety. 2019. Available from: https://www.actascientific.com/ASNH/ASNH-04-0593.php [Accessed: July 20, 2021]

[57] Crystal Chocolatier. 4 Steps to Maintain Good Hygiene in a Chocolate Factory. 2013. Available from: https://www.crystalchocolatier.com/4-steps-maintain-good-hygiene-chocolate-factory [Accessed: July 27, 2021]

[58] Johnston I. Food Safety Protocols for Chocolate Companies. 2021. Available from: https://smallbusiness.chron.com/food-safety-protocols-chocolate-companies-74340.html [Accessed: July 25, 2021]

[59] Aliw. Chocolate production hazards – HACCP Easter everyone!. 2018. Available from: https://www.vwa.co.uk/blog/news/5970/ [Accessed: July 25, 2021]

Section 3

Packaging

Chapter 5

Pectin and Its Applicability in Food Packaging

Heba Younis, Guohua Zhao and Hassan Abdellatif

Abstract

Food packaging based on plastic films made from nonrenewable resources often causes environmental problems after disposal. Recently, researchers are increasingly focusing on alternative materials to reduce the use of nonbiodegradable and nonrenewable films. Generally, biomaterials are nontoxic, biocompatible, and renewable always presents reasonable film-forming ability. Thus, they are important for food safety, where undesired chemical compounds might migrate from chemicals migrate from the plastic packaging materials into foods. Pectin (PEC), as a natural carbohydrate polymer, belongs to the anionic heteropolysaccharide family and is often extracted from various residues from plant food processing, such as apple and citrus pomaces. The pectin molecules are highly branched with a backbone α-(1–4) linked D galacturonic acid. Among the naturally derived carbohydrate-based biopolymers, pectin was considered a promising substrate in fabricating edible films due to its diverse advantages, such as perfect film-forming ability, evidenced bioactivity, easy availability, and excellent quality biodegradability and biocompatibility, nontoxicity, and low cost. Pectin-based films have excellent oxygen barrier capacity and extend the shelf life for different fruits. The properties of pure pectin films can enhance through combination with other polymers or nanoparticles/fibers.

Keywords: pectin, food packaging, edible films, biopolymers

1. Introduction

Packaging plays a fundamental role in all our daily lives purposes, especially in food industries for food products (fruit and vegetables) during post-harvest and processing due to their different uses, such as; 1) Separate the food from the surrounding environment, 2) Preserving food quality, and safety, 3) Protecting from the spoilage factors, 4) Maintaining the nutritional value of products, 5) Extending shelf life, and 6) Provides information about the products for the consumers [1–3]. The packaging process was defined as the art, science, and technology to deliver the different goods to the consumers at economical prices, securely, and high quality [4].

Fossil fuels (include coal, petroleum, natural gas, oil shales, bitumens, tar sands, and heavy oils) are considered the main source for commercial polymers used in food packaging and producing plastic packaging [2, 5]. This is due to several reasons, such as ease of processing, low cost, lightweight, high energy effectiveness, and flexibility. The use of plastic packaging increases worldwide, owing to increasing population growth, which leads to increasing demand for processed food [6–8]. The plastic packaging production reached 380 million metric tons worldwide, approximately 60% of all plastic packaging used for foods and beverages, and the remainder covers nonfood applications [9].

Despite the different advantages of plastic polymers, it considered highly hazardous due to containing a large variety of chemical additives which using during manufacturing; thus, it has a severe threat to consumer's health and environmental pollution [10–15] such as 1) Produce a large of wastes which can be toxic to biological life [16–18]. 2) Migration might result in accumulation of undesired chemical compounds from packaging materials to foods, such as Bisphenol A (BPA) [12, 19]. 3) Lead to toxic and harmful effects on wild and human life because it is nonbiodegradable and not reusable [6, 20, 21].

In 2015, global plastic waste volume reached around 6.3 billion metric tons and is expected to increase to 12 billion metric tons by 2050 [22].

Therefore the food industry has classified the conventional plastic packaging material as a source of pollution and social concerns due to its nonbiodegradability and poor waste management, due to the accumulation of huge amounts of plastic waste in the environment, and also rapid depletion of fossil reserves and increases in the cost of petroleum, the food packaging industry toward the development and application of eco-friendly materials and biodegradable materials [23].

Various technologies have been used in food packaging to preserve the fruit and vegetable's quality and safety and prolong their shelf life, such as ultraviolet irradiation, ozonation, changed atmosphere, and biodegradable films [24]. Edible packaging is considered a promising alternative and received much attention to tackle the plastics packaging problems [25–27].

Production of environmental-friendly packaging such as biodegradable and renewable films represents an interesting alternative to conventional polymers. Polysaccharides and proteins address those requirements because of their desirable film-forming properties, and, as most of them are also edible, they can be used as edible films and coatings. Edible films and coatings can extend the shelf life of food products by improving the mechanical properties and reducing the transfer of moisture, lipids, flavors, or gases between the food and the surrounding environment [17]. Edible coatings modify the atmosphere around fruit and/or vegetables, altering oxygen levels inside the fruit, retarding the production of ethylene and thus, limiting the physiological decay of fruits; This also reduces the ripening-induced quality degradation in terms of texture or loss of bioactive compounds during storage [28].

Edible coating materials efficiency and performance depend on i) Properties of coating materials (type, concentration, viscosity). ii) Methods of coating (dipping, spraying, and dripping) [29–31]. A coating with excellent barrier and poor mechanical properties, bad flavor, or high cost will not be of commercial interest for using, and selecting a proper coating for the fruits also is not easy. So, it is necessary to study the film's physicochemical and structural characteristics [32, 33].

Pectin as edible films/coating used to increase food quality and extend the shelf life of food products which is considered an alternative to packaging materials from synthetic polymers, thus preventing environmental pollution from nonbiodegradable plastic materials [34].

Younis and Zhao [35] reported that pectin polymers are considered promising polysaccharide biomaterials in developing eco-friendly films due to their film-forming, biodegradable, and nontoxic characteristics.

2. Pectin

Pectin (PEC) is a natural polymer and complex anionic heteropolysaccharides [36–39], consisting mainly of α-(1-4) linked D-galacturonic acid residues and neutral sugars, which are partially esterified with methyl alcohol or acetic acid at the carboxylic acid [37, 40].

Pectin is a family of plant polysaccharides accounting for up to 35% of primary cell walls in certain species. It has been considered the most structurally complex polysaccharide in nature. The term "pectin", in fact, describes a group of oligosaccharides and polysaccharides that share common features but are highly diverse in their fine structure, except that they all comprise at least 65% galacturonic acid (GalA), which is linked at the O-1 and O-4 position [41].

Zhou et al. [36] noted that the pectin polymer consists of three polysaccharide domains: Homogalacturonan (HGA), Rhamnogalacturonan-I (RG-I), and Rhamnogalacturonan-II (RG-II), where Homogalacturonan (HG) considered the most abundant cell wall pectic polysaccharide, about 50–90% of total pectin, meanwhile rhamnogalacturonan I (RG-I) is the second most abundant type comprising between 20% and 35% of total pectin [42, 43].

Pectin is a white, amorphous, and colloidal carbohydrate of high molecular weight occurring in ripe fruits, especially in apples, currants, etc., and used in fruit jellies, pharmaceuticals, and cosmetics for its thickening and emulsifying properties and ability to solidify to a gel. These properties and applications have put pectin in the biopolymers market with great potential for future developments [44].

2.1 Pectin sources

Several studies have documented that pectin is one of the most abundant polysaccharides in the primary cell wall and middle lamella of all plant tissues [39, 45]; the pectin production comes from two ways i) Commercial pectin comes from citrus fruit peels [46, 47], apple pomace [47, 48] pomegranate peels [49, 50], mango peel [3, 51], lemon peel [52], sugar beet pulp and potato pulp [3, 51]; ii) Noncommercial pectin obtained from cocoa husks [53], mulberry branch bark [54], peach pomace [42], sisal waste [42], pumpkin [51], banana peel [55, 56] watermelon rind, and soy hull [38].

2.2 Pectin extraction

There are different techniques for pectin extraction which is considered highly efficient and eco-friendly such as conventional solvent extraction based on stirring and heating; microwave-assisted extraction (MAE); ultrasound-assisted extraction [57, 58]; subcritical water extraction, and enzyme-assisted extraction [59], each of them has advantages and disadvantages.

Wicker et al. [60] noted that the pectin polymer could easily be extracted using different acids by conventional heating extraction. It consumes more time, the hot solution leads to pectin degradation, produces a large volume of effluent, and causes environmental pollution concerns. The MAE method is considered an excellent alternative to conventional extraction methods and offers significant advantages 1) Requires shorter processing times and less solvent, 2) Higher extraction rates, 3) Low in cost [52, 59].

The microwave-assisted extraction method (MAE) has been reported as the preferred extraction method of pectins from natural sources such as dragon fruit peels, bagasse, and pomace obtained from Mexican lime fruit, pomelo, mango, and papaya peels under different operation conditions. MAE conditions are dependent on different factors, such as microwave power, pH, time, and S:L ratio. MAE methods show significant advantages over conventional extraction techniques, such as reducing the amount of the extraction solvents, low energy consumption, high recoveries, good reproducibility, short extraction times (minutes rather than hours), and minimal sample manipulation [44].

The percentage of galacturonic acid group of pectin is esterified using methyl, or acetyl groups are termed the degree of esterification [61]. According to the degree of esterification (DE), PEC is commonly categorized into: High methoxyl pectin (HMP) (DE > 50%), and low methoxyl pectin (LMP) (DE < 50%) [57, 62, 63]. The degree of methyl-esterification can be defined as the percentage of carboxyl groups esterified with methanol of the pectin or the number of moles of methanol per 100 moles of galacturonic acid [64].

Pectins are usually extracted by hot dilute mineral acids at pH 1.5–2.5, taking 2–4 h with further precipitation with ethanol or isopropyl alcohol, separation to remove impurities, drying, grinding, and blending with other additives. The extracted pectins can be categorized into two major types, depending on their DM degree, and different factors like pH, temperature, time, and solvent: liquid (S:L) ratio are usually studied to optimize the extraction conditions [44].

Valdés et al. [44] showed that the different natural pectins extracted by solvent extraction methods; such as tomato, banana peel, and sugar beet, the tomato peel pectin can be extracted through two steps by ammonium oxalate and oxalic acid at 90°C (24 and 12 h) and pectin yield extraction (PYE) of about 32.0%, while Sugar beet pulp pectin produced by citric acid, pH 1, 166 min at 99°C, 1:20 g·mL^{-1} 23.95%. Banana peel using a citric acid solution with pH 2.0 at 160 min, 87°C, 1:20 g·mL^{-1} give a lower PYE 13.89%, the PYE can be increased to 16.54% using citric acid and HCl, pH 1.5, 4 h, 90°C, but the pectin from mango peel can extract using sulfuric acid in water with pH 1.5, and 2.5 h at 90°C which give a higher extraction yields about >70%.

The HMP and LMP pectin have different physicochemical characteristics and thus different applications for each of them [53]; therefore, the degree of esterification plays an important role and is a good parameter for the biochemical, physical properties of pectin applications [16, 65]. Cho et al. [66] reported that the HMP pectin could be converted into LMP using two ways: 1) Chemical de-esterification by alkali; 2) Enzymatic treatment by pectin methylesterase. The LM-pectin is mainly produced by the chemical de-esterification of HM-pectin with acid, alkali, and alcoholic/aqueous ammonia [67].

Depending on the raw material and the extraction method, pectins have various molecular weights, structures, and functional groups. These characteristics influence and determine the techno-functional properties of pectin molecules, such as gelling properties [43].

However, the industrial production of LM-pectin by enzymatic de-esterification of HM-pectin remains a challenge. For successful industrial application, it is necessary to enhance pectin methylesterase (PME) productivity by genetic modification and optimizing culture conditions [67].

2.3 Properties and advantages of pectin

Several studies have reported that the pectin polymers have several properties and advantages such as renewability, biodegradability, and biocompatibility [20, 68, 69], amphiphilic properties (hydrophobic/hydrophilic nature) [48, 49], pectin-based films showing high tensile strength, and water resistance [70], but the applications of pectin is limited due to their poor chemo-physical properties and has poor water barrier properties if it used alone [3, 71]. Younis et al. [34] noted that the pure pectin films encountered several defects when used alone without any other polymers, such as soluble in water, high water permeability, poor waterproof, and unsatisfactory mechanical properties. These defects heavily limited the application of pectin films, especially in an atmosphere with high humidity or to package high-moisture foods.

2.4 Pectin applications

Pectin is considered a good source for human health, where its consumption by large volume in our life due to its ability to decrease blood cholesterol levels [37, 42, 51], and has been widely used in the food and nonfood industries; 1) Food and beverage industries, as edible films [3], gelling and thickener, texturizer [39, 49, 72], emulsifier agent [38, 49], colloidal stabilizer in food products such as jams, yogurt drinks, dairy products, and ice cream [51, 72, 73], also carrier polymer for the encapsulation of food ingredients [74]. 2) Pharmaceutical and biomedical applications [51, 75], including drug delivery, gene delivery [69], cosmetics [39], wound healing, and tissue engineering hydrogels as carriers for tissue regeneration [68, 76].

3. Pectin in food industry and food packaging

For the food industries, several studies showed that the pectin has an excellent potential to be an edible film for food packaging and biocomposite materials [3, 20]; the presence of methoxyl groups in the pectin structure leads to an increase in the hydrophobic nature of pectin molecules that help the pectin to use in different food applications such as emulsifier agent and stabilizer [48]. The gels from the HM pectin are formed under sugar or acidic conditions, while from LM pectin formed through ionotropic gelation with low-valence ions, such as calcium ions, due to its containing a large number of ionizable carboxylic groups; therefore, they have a strong affinity for calcium ions [68], where the carboxyl group linked with the D-galacturonic acid residue exists in two essential forms: 1) carboxylate salt; 2) neutral methoxylated or ester [77]. Natural hydrocolloid pectin is widely utilized for thickening, stabilization, and encapsulation in the food & beverage, cosmetic, and pharmaceutical industries [67].

Pectins are hydrocolloids typically used as thickening and gelling agents, e.g., sauces and jams [43]. The use of pectin as a thickening and gelling agent, for example, sauces and jams, also gaining interest as a potential food emulsifier of natural origin; it can be used as a multifunctional ingredient in different food applications [78]. Einhorn-Stoll et al. [79] found that citrus pectin can be used as a thickening and gelling agent in a wide range of foods. At the same time, Liu et al. [80] showed that the pectin produced from sugar beet pulp (SBP) is the most prominent for stabilizing oil-in-water (O/W) emulsions and has received increasing attention in the food industry as a polysaccharide-based emulsifier.

The most recent trends in the field of pectins coating applications include; the shelf-life extension of fresh-cut highly perishable food, the application of pectin coatings as a pre-frying treatment to reduce the oil consumption in deep-fat fried products, and the use of pre-dried treatments to improve the retention of nutrients and quality characteristics of dehydrated and lyophilized food [44].

Due to its biodegradability, biocompatibility, edibility, chemical and physical properties, pectin is considered an applicable polymeric matrix for elaborating edible films intended as active food packaging [40]. Pectin-based biofilms could extend the shelf life of avocado fruits over a month compared to uncoated fruits by decreasing the oxygen absorption and thus delaying texture and color change [81].

Sucheta et al. [82] noticed that the performance of the commercial pectin films was affected by the incorporation of cornflour and beetroot; significant differences were found in tensile strength (TS), water solubility (WS), and the thickness film, where the TS increased when the pectin added to cornflour and beetroot from 1.36 to 7.47 MPa and 3.79 MPa. The WS decreased from 97.8% to 70.7%, also decreased film thickness from 0.24 to 0.06 mm and 0.09 mm, respectively. The effect of pectin with cornflour and beetroot as a coating has been studied of tomatoes fruits by

Sucheta et al. [28] and find that the best coating for tomatoes was pectin/cornflour, it showed a significantly reduced the weight loss %, also delayed the respiration rate with retention of internal quality during the storage period (30 days) improved the shelf life of tomatoes and showed minimum shrinkage during the end of storage.

Meerasri and Sothornvit [83] find that the pectin polymer could provide a combination with a plasticizer (glycerol) and bioactive compounds (Gamma-aminobutyric acid [GABA]) and affect its performance; GABA incorporation showed an excellent enhancement in the barrier water than the glycerol addition where the WVP for the pectin decreased from 3.59×10^{-10} to $2.54 \times 10\ g^{-10}\ g{\cdot}m^{-1}{\cdot}s^{-1}{\cdot}Pa^{-1}$ (pectin/GABA). In contrast, the glycerol incorporation led to increasing the WVP to $3.73 \times 10\ g^{-10}\ g{\cdot}m^{-1}{\cdot}s^{-1}{\cdot}Pa^{-1}$, the mechanical properties of pectin films were significantly affected by glycerol and GABA incorporation, the TS for the pure pectin film 6.41% decreased to 2.14 and 2.17% for glycerol and GABA, respectively, while the Elongation at break (EB) increased from 8.78 (pectin film) to 15.11 and 27.99 MPa, respectively.

Sucheta et al. [28] studied the effect of pectin-corn flour-based coating and observed that it was the best treatment that reduced weight loss and decay, delayed respiration with retention of biochemical quality, and improved the shelf life of tomatoes.

Chitosan/poly (vinyl alcohol)/pectin ternary film was prepared by solution casting method; the films showed the antimicrobial activity against different pathogenic bacteria, such as *Escherichia coli*, *Staphylococcus aureus*, *Bacillus subtilis*, *Pseudomonas spp* [84].

Lorevice et al. [85] studied the effect both of HMP (High methoxyl pectin) and LMP (Low methoxyl pectin) with chitosan nanoparticles (DD = 94%) as edible films using water as solvent and find that the addition of chitosan nanoparticles (CSNPs) to pectin films did not cause remarkable visual changes but improved the mechanical and physical properties when compared with pure pectin films as a result of stronger interactions between the polymer matrices; The pure LMP films recorded a lower values than HMP films for the thickness and the mechanical properties, where the thickness, Elongation at break (EB) and Tensial strength (TS) of pectin films was remarkably increased after the addition of CSNPs; For the TS increased from 30.81 MPa to 46.95 MPa for HMP pectin films; whereas LMP films increased from 26.07 MPa to 58.51 MPa, the EB recorded a higher value for the LMP/CSNP 2.91% compared with pure LMP pectin (0.94%), the enhanced in the EB values of LMP films incorporated with CSNPs may be explained by the higher number of interactions between this pectin and CSNP surface, where it has more carboxylic groups that interact with CSNP amine groups.

The thermal properties performed by differential scanning calorimetry (DSC) analysis and noted that the heat absorbed (ΔH, change in enthalpy) by pectin films was larger than by pectin powder and PEC/CSNPs films where it increases from $401.6 \rightarrow 444.7 \rightarrow 547.9\ J{\cdot}g^{-1}$ for HMP powder, HMP films and HMP/CSNP film, respectively. Regarding the endothermic peak temperatures (EPT) property, it decreased when CSNPs were added from $121.8 \rightarrow 105.1 \rightarrow 105.9$°C for HMP powder, HMP films, and HMP/CSNP film, respectively. Also, for the LMP, the EPT decreased from $108.9 \rightarrow 97.8$°C, for LDM pectin powder, LMP films, respectively, while by the CSNP addition, the temperature increased to 102.1°C. this can be attributable to reduced natural hydration degrees of pectin films due to CSNP presence, the lower hydration degrees may result in lower chain mobility, the lower EPTs, and higher ΔH values suggest that CSNPs were occupying water sites within pectin matrices and led to absorb more thermal energy than pure pectin films.

For the barrier properties, the addition of CSNPs into the LMP films made a significant difference and decreased the water vapor transmission rate and permeance property from 118.67 to 96.56 $g{\cdot}h^{-1}{\cdot}m^{-2}$ and 81.04 to 54.06 $g{\cdot}kPa^{-1}{\cdot}h^{-1}{\cdot}m^{-2}$, respectively, the effect of reducing permeance caused by CSNP presence was more remarkable than in HMP since LMP has a higher content of free hydroxyl groups than HMP [85].

Ngo et al. [86] evaluated the films from pectin (2% w/v) and nano chitosan (2% w/v) with different ratios (100:0, 75:25, 50:50, 25:75, and 0:100 w/w) and find that the blending ratio 50:50 increased the tensile strength and decreased the water solubility, water vapor permeability, and oxygen permeability; Also, these films showed antimicrobial activity against *C. gloeosporioides*, *S. cerevisiae*, *A. niger*, and *E. coli*.

Ranganathan et al. [87] noted that pectin polymer has the great potential to form a polymer composite and makes it flexible film due to its anionic property, whereas; Tsai et al. [88] found that the addition of pectin into chitosan with a high degree of deacetylation increased the viscosity of the matrix solution, and making it more difficult to prepare homogeneous solutions; also mentioned if the densities of positive charges on the polycations and negative charges on polyanions are not equitable, the solution will be water-soluble and form homogeneous.

Chen et al. [89] showed that when chitosan was mixed with pectin, the mechanical properties and hydrophilicity increased for the chitosan/pectin films, where they prepared the chitosan/pectin membranes successfully via using a freeze-gelation method and proved that the combination of pectin into chitosan could significantly enhance the properties of the film including and increased the tensile strength from 8 $N{\cdot}g^{-1}$ (CH films) to 32 $N{\cdot}g^{-1}$ (CH: PEC films) also decreased the water contact angle from 85° (CH films) to 45° (CH: PEC films); therefore it can be used as a carrier for food or drug.

Norcino et al. [16] used the low molecular weight chitosan with the degree of deacetylation of 80% and pectin from the peel of citrus fruits to prepare different CH/PEC films by solution blending and observed the following; The contact angle values were clearly visible for pure CH film, this can be related with the sorption of water on the surface, and film swelling, followed by water absorption, the water contact angle for CH/ PEC blends 50/50 and 25/75 recorded a higher value (87°) while for the pure CH, PEC, and CH/PEC 75/25 films were $\sim 64^\circ$, 75°, and 74° respectively and remained practically constant even after 60 s. The ionic cross-linking effect restricted the CS/PEC blend film swelling and reduced their water sorption.

For the mechanical properties, the tensile strength increased for the CH/PEC blend films. It is evident that the tensile strength of CH/PEC blends presented a large positive deviation from addition law; While pure CH and PEC films showed lower values for the tensile strength 41 and 48 MPa, respectively, while it increased for CH/PEC blend 75/25 was 70 MPa and reached 81 MPa for CH/PEC blends 50/50 and 25/75.

Ionic crosslinking between the chitosan ammonium groups (NH_3^+) and pectin carboxylate groups (COO−) and thus playing an important role on the physical properties of CH/PEC polyelectrolyte based-films where the electrostatic interactions were responsible for increasing water resistance and mechanical properties of chitosan/pectin films, which allows modulating their thermo-mechanical and other physical properties for specific applications like medicine, agriculture, and food packaging [16].

Younis and Zhao [35] prepared the different films from chitosan (deacetylation degree 67.9%) and high methoxyl apple pectin by *casting* method. They noted that the tensile strength increased from 1.22 MPa (CH) to 5.06 MPa (CH/ PEC), while the elongation at break for the chitosan did not affect by the pectin incorporation; For the water vapor permeability WVP values of the PEC, CH and blend films recorded a significant difference where the WVP value of PEC film ($5.53 \times 10\ g^{-12}\ g\ cm^{-1}{\cdot}s^{-1}{\cdot}Pa^{-1}$) was much higher than that of PEC/CH film ($3.91 \times 10\ g^{-12}\ g\ cm^{-1}{\cdot}s^{-1}{\cdot}Pa^{-1}$). Still, the blending does not influence the barrier properties. The more porous structure of PEC film contributes to its higher WVP;

the significant reduction in WVP makes the films more suitable for food packaging. Meanwhile, the surface morphology for the films was observed by scanning electron microscopy (SEM), and numerous small pores were evenly distributed in PEC film while big caves were unevenly scattered in CH film; by incorporating the pectin into chitosan, these cavities disappeared, as shown in **Figure 1**.

Chetouani et al. [90] prepared edible films by *casting* method from apple pectin (DE, 70–75%) and chitosan (viscosity of 800,000 cps) and compared it with chitosan/oxidized pectin by sodium meta periodate ($NaIO_4$); the results showed the following; The X-ray diffraction, as shown in **Figure 2** showed that the chitosan films exhibit broad diffraction peaks observed at (2θ) = 9.8° and 19.9°, which are typical fingerprints of semi-crystalline chitosan. The pure pectin films showed sharp peaks centered at 9.2°, 12.8°, 18.5°, 28.1°, and 40.1° and were considered a semi-crystalline material. After the pectin's oxidation process, the X-ray diffraction analysis (XRD) is clearly different from that of pure pectin. The peaks at 16.5°, 24.1°, 27.5°, and 33.5° indicate less crystalline the oxidized pectin (OPEC).

From the above results, the films confirmed that CH's crystallinity is drastically decreased with the addition of PEC or oxidized PEC. Also, the addition of PEC to CH makes the material amorphous; this could be due to the interacts of chitosan with pectin through inter-molecular hydrogen bonding.

Chetouani et al. [90] noted that the pectin, CH, CH/pectin, and CH/oxidation pectin films do not affect the Gram-negative bacteria (*P. aeruginosa, E. coli*). In contrast, Gram-positive bacteria (*B. subtilis and S. aureus*) are more sensitive to the films. Also, there is a strong interaction between chitosan and oxidized pectin films, which influences the thermal decomposition; using these types of films can improve the antibacterial activity of chitosan, which demonstrates them to be promising materials for food packaging and biomedical applications.

Baron et al. [91] reported that the chitosan CH (blue crab waste) and PEC (orange peel) can produce edible films by *casting* method; they noted that the addition of high concentrations of chitosan with pectin polymer in the films formulations (75:25 ratio) led to producing films with lower solubility and lower moisture content 10.2% and 15.8% respectively, compared with the values of pure pectin; also reduced the water vapor permeability from $1.06 \times 10\ g^{-15}\ g{\cdot}m^{-1}{\cdot}s^{-1}{\cdot}Pa^{-1}$ (pectin films) to $0.99 \times 10\ g^{-15}\ g{\cdot}m^{-1}{\cdot}s^{-1}{\cdot}Pa^{-1}$ (75 CH:25 PEC), when the quantity of protonated free amino groups (NH_3^+) in chitosan and the free anionic groups present in the pectin becomes higher in the films, the degree of swelling increased from 14.3% (PEC films) to 15.5% (CH: PEC), while for the mechanical properties, increasing chitosan proportion in biopolymers blend increased the tensile strength (TS) and become more elastic and flexible from 17 MPa (PEC) to 23 MPa (75 CH:25 PEC) also reduced deformability (EB) 37.7–27%, respectively.

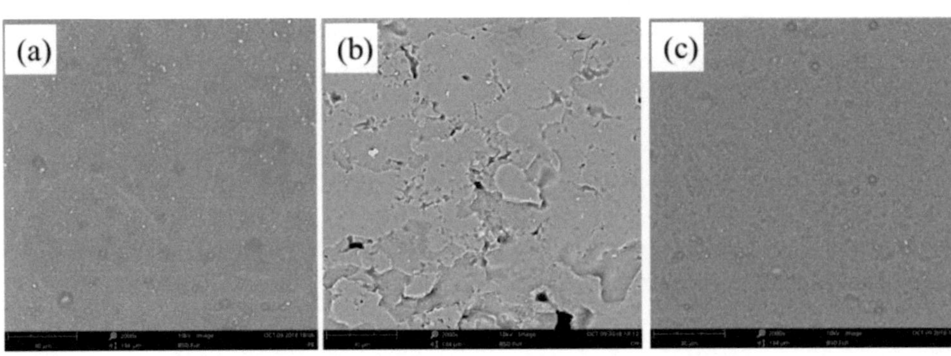

Figure 1.
SEM micrographs of the films surface a) pectin, b) chitosan, c) pectin/chitosan films [35].

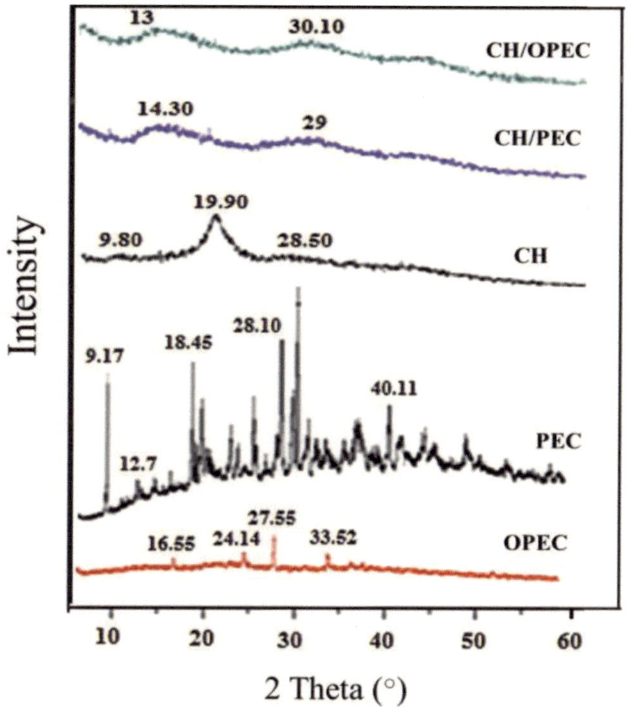

Figure 2.
X-ray diffractions for the films of pectin (PEC), oxidized pectin (OPEC), and chitosan (CH).

4. Conclusion

Today, the food industry's important challenges are developing environmentally friendly, green, and intelligent food packaging materials to avoid environmental problems caused by synthetic polymer packaging materials. Edible coating of biodegradable materials is considered the most trusted technique for storing fruit and vegetables, an excellent alternative to chemically synthesized preservatives, more effective in prolonging various agricultural products' shelf life, and inhibiting disease growth on the fruit surface. Biodegradable materials have several characteristics, including biocompatibility, biodegradability, renewability, non-toxicity. Therefore, in recent times the researchers have been focused on edible packaging such as pectin polymers. Pectin is a promising material that can avoid nonbiodegradable polymers problems when applied as an edible coating for different fruits due to its different properties.

Conflict of interest

The authors declare no conflict of interest.

Author details

Heba Younis[1], Guohua Zhao[2] and Hassan Abdellatif[1*]

1 Faculty of Agriculture, Agricultural Engineering Department, Cairo University, Giza, Egypt

2 College of Food Science, Southwest University, Chongqing, China

*Address all correspondence to: hassan_elshemy2009@cu.edu.eg

References

[1] Ma D, Jiang Y, Ahmed S, Qin W, Liu Y. Physical and antimicrobial properties of edible films containing Lactococcus lactis. International Journal of Biological Macromolecules. 2019;**141**:378-386

[2] Otoni CG, Moura MR, Aouada FA, Camilloto GP, Cruz RS, Lorevice MV, et al. Antimicrobial and physical-mechanical properties of pectin/papaya puree/cinnamaldehyde nanoemulsion edible composite films. Food Hydrocolloids. 2014;**41**:188-194

[3] Šešlija S, Nešić A, Ružić J, Kalagasidis Krušić M, Veličković S, Avolio R, et al. Edible blend films of pectin and poly(ethylene glycol): Preparation and physico-chemical evaluation. Food Hydrocolloids. 2018;**77**:494-501

[4] Priyadarshi R, Rhim J-W. Chitosan-based biodegradable functional films for food packaging applications. Innovative Food Science & Emerging Technologies. 2020;**62**:102346

[5] Pacheco N, Naal-Ek MG, Ayora-Talavera T, Shirai K, Román-Guerrero A, Fabela-Morón MF, et al. Effect of bio-chemical chitosan and gallic acid into rheology and physicochemical properties of ternary edible films. International Journal of Biological Macromolecules. 2019;**125**: 149-158

[6] Ahmed J, Arfat Y, Bher A, Mulla M, Jacob H, Auras R. Active chicken meat packaging based on polylactide films and bimetallic Ag–Cu nanoparticles and essential oil. Journal of Food Science. 2018;**83**(5):1299-1310

[7] Sartori T, Feltre G, do Amaral Sobral PJ, Lopes da Cunha R, Menegalli FC. Properties of films produced from blends of pectin and gluten. Food Packaging and Shelf Life. 2018;**18**:221-229

[8] Gigli M, Negroni A, Zanaroli G, Lotti N, Fava F, Munari A. Environmentally friendly PBS-based copolyesters containing PEG-like subunit: Effect of block length on solid-state properties and enzymatic degradation. Reactive and Functional Polymers. 2013;**73**(5):764-771

[9] Groh KJ, Backhaus T, Carney-Almroth B, Geueke B, Inostroza PA, Lennquist A, et al. Overview of known plastic packaging-associated chemicals and their hazards. Science of The Total Environment. 2019;**651**:3253-3268

[10] Atarés L, Chiralt A. Essential oils as additives in biodegradable films and coatings for active food packaging. Trends in Food Science & Technology. 2016;**48**:51-62

[11] Mahcene Z, Khelil A, Hasni S, Akman PK, Bozkurt F, Birech K, et al. Development and characterization of sodium alginate based active edible films incorporated with essential oils of some medicinal plants. International Journal of Biological Macromolecules. 2020;**145**:124-132

[12] Rodríguez FJ, Galotto MJ, Guarda A, Bruna JE. Modification of cellulose acetate films using nanofillers based on organoclays. Journal of Food Engineering. 2012;**110**(2):262-268

[13] Espitia PJP, Du W-X, Avena-Bustillos RJ, Soares NFF, McHugh TH. Edible films from pectin: Physical-mechanical and antimicrobial properties - A review. Food Hydrocolloids. 2014;**35**:287-296

[14] Tahir HE, Xiaobo Z, Mahunu GK, Arslan M, Abdalhai M, Zhihua L. Recent developments in gum edible coating applications for fruits and vegetables preservation: A review. Carbohydrate Polymers. 2019;**224**:115141

[15] Pellá MCG, Silva OA, Pellá MG, Beneton AG, Caetano J, Simões MR, et al. Effect of gelatin and casein additions on starch edible biodegradable films for fruit surface coating. Food Chemistry. 2020;**309**:125764

[16] Norcino LB, de Oliveira JE, Moreira FKV, Marconcini JM, Mattoso LHC. Rheological and thermo-mechanical evaluation of bio-based chitosan/pectin blends with tunable ionic cross-linking. International Journal of Biological Macromolecules. 2018;**118**:1817-1823

[17] Khodaei D, Oltrogge K, Hamidi-Esfahani Z. Preparation and characterization of blended edible films manufactured using gelatin, tragacanth gum and, Persian gum. LWT. 2020;**117**: 108617

[18] Qin Y, Liu Y, Zhang X, Liu J. Development of active and intelligent packaging by incorporating betalains from red pitaya (Hylocereus polyrhizus) peel into starch/polyvinyl alcohol films. Food Hydrocolloids. 2020;**100**:105410

[19] Homez Jara A, Daza L, Aguirre D, Muñoz A, Solanilla Duque JF, Váquiro H. Characterization of chitosan edible films obtained with various polymer concentrations and drying temperatures. International Journal of Biological Macromolecules. 2018;**113**:1233-1240

[20] Gurram R, Souza Filho P, Taherzadeh M, Zamani A. A solvent-free approach for production of films from pectin and fungal biomass. Journal of Polymers and the Environment. 2018;**26**:4282-4292

[21] Nešić A, Gordić M, Davidović S, Radovanović Ž, Nedeljković J, Smirnova I, et al. Pectin-based nanocomposite aerogels for potential insulated food packaging application. Carbohydrate Polymers. 2018;**195**:128-135

[22] Geyer R, Jambeck JR, Law KL. Production, use, and fate of all plastics ever made. Science Advances. 2017;**3**(7):e1700782

[23] Haghighi H, Licciardello F, Fava P, Siesler H, Pulvirenti A. Recent advances on chitosan-based films for sustainable food packaging applications. Food Packaging and Shelf Life. 2020;**26**:100551

[24] Tóth A, Halász K. Characterization of edible biocomposite films directly prepared from psyllium seed husk and husk flour. Food Packaging and Shelf Life. 2019;**20**:100299

[25] Liu J, Liu S, Wu Q, Gu Y, Kan J, Jin C. Effect of protocatechuic acid incorporation on the physical, mechanical, structural and antioxidant properties of chitosan film. Food Hydrocolloids. 2017;**73**:90-100

[26] Lima M, Carneiro LC, Bianchini D, Dias ARG, Zavareze ER, Prentice C, et al. Structural, thermal, physical, mechanical, and barrier properties of Chitosan films with the addition of Xanthan gum. Journal of Food Science. 2017;**82**(3):698-705

[27] Khazaei N, Esmaiili M, Djomeh ZE, Ghasemlou M, Jouki M. Characterization of new biodegradable edible film made from basil seed *(Ocimum basilicum L.)* gum. Carbohydrate Polymers. 2014;**102**: 199-206

[28] Sucheta, Chaturvedi K, Sharma N, Yadav SK. Composite edible coatings from commercial pectin, corn flour and beetroot powder minimize post-harvest decay, reduces ripening and improves sensory liking of tomatoes. International Journal of Biological Macromolecules. 2019;**133**:284-293

[29] Jiang Y, Yu L, Hu Y, Zhu Z, Zhuang C, Zhao Y, et al. Electrostatic spraying of chitosan coating with

different deacetylation degree for strawberry preservation. International Journal of Biological Macromolecules. 2019;**139**:1232-1238

[30] Thakur R, Pristijono P, Scarlett CJ, Bowyer M, Singh SP, Vuong QV. Starch-based edible coating formulation: Optimization and its application to improve the postharvest quality of "Cripps pink" apple under different temperature regimes. Food Packaging and Shelf Life. 2019;**22**:100409

[31] Mannozzi C, Tylewicz U, Chinnici F, Siroli L, Rocculi P, Dalla Rosa M, et al. Effects of chitosan based coatings enriched with procyanidin by-product on quality of fresh blueberries during storage. Food Chemistry. 2018;**251**:18-24

[32] Olivas G, Dávila-Aviña J, Salas-Salazar NA, Molina FJ. Use of edible coatings to preserve the quality of fruits and vegetables during storage. Stewart Postharvest Review. 2008;**4**:1-10

[33] Tapia-Blácido DR, Maniglia B, Martelli TM. Transport phenomena in edible films. Polymers for Food Applications. 2018:149-192

[34] Younis HGR, Abdellatif HRS, Ye F, Zhao G. Tuning the physicochemical properties of apple pectin films by incorporating chitosan/pectin fiber. International Journal of Biological Macromolecules. 2020;**159**:213-221

[35] Younis HGR, Zhao G. Physicochemical properties of the edible films from the blends of high methoxyl apple pectin and chitosan. International Journal of Biological Macromolecules. 2019;**131**:1057-1066

[36] Zhou W, Liu W, Zou L, Liu W, Liu C, Liang R, et al. Storage stability and skin permeation of vitamin C liposomes improved by pectin coating. Colloids and Surfaces B: Biointerfaces. 2014;**117**:330-337

[37] Xu Y, Zhang L, Bailina Y, Ge Z, Ding T, Ye X, et al. Effects of ultrasound and/or heating on the extraction of pectin from grapefruit peel. Journal of Food Engineering. 2014;**126**:72-81

[38] Hosseini SS, Khodaiyan F, Ms Y. Optimization of microwave assisted extraction of pectin from sour orange peel and its physicochemical properties. Carbohydrate Polymers. 2016;**140**:59-65

[39] Pasini Cabello SD, Takara EA, Marchese J, Ochoa NA. Influence of plasticizers in pectin films: Microstructural changes. Materials Chemistry and Physics. 2015;**162**:491-497

[40] Yeddes W, Djebali K, Aidi Wannes W, Horchani-Naifer K, Hammami M, Younes I, et al. Gelatin-chitosan-pectin films incorporated with rosemary essential oil: Optimized formulation using mixture design and response surface methodology. International Journal of Biological Macromolecules. 2020;**154**:92-103

[41] Wusigale, Liang L, Luo Y. Casein and pectin: Structures, interactions, and applications. Trends in Food Science & Technology. 2020;**97**:391-403

[42] Santos JDG, Espeleta AF, Branco A, de Assis SA. Aqueous extraction of pectin from sisal waste. Carbohydrate Polymers. 2013;**92**(2):1997-2001

[43] Saavedra Isusi GI, Bindereif B, Karbstein HP, van der Schaaf US. Polymer or microgel particle: Differences in emulsifying properties of pectin as microgel or as individual polymer chains. Colloids and Surfaces A: Physicochemical and Engineering Aspects. 2020;**598**:124793

[44] Valdés A, Burgos N, Jiménez A, Garrigós CM. Natural pectin polysaccharides as edible coatings. Coatings. 2015;**5**(4):865-886

[45] Guerreiro AC, Gago CML, Faleiro ML, Miguel MGC, Antunes MDC. The effect of edible coatings on the nutritional quality of 'Bravo de Esmolfe' fresh-cut apple through shelf-life. LWT. 2017;**75**:210-219

[46] Seixas F, Turbiani F, Salomao P, Souza R, Gimenes M. Biofilms composed of alginate and pectin: Effect of concentration of crosslinker and plasticizer agents. Chemical Engineering Transactions. 2013;(32):1693-1698

[47] Willats WGT, Knox JP, Mikkelsen JD. Pectin: New insights into an old polymer are starting to gel. Trends in Food Science & Technology. 2006;**17**(3):97-104

[48] Mungure TE, Roohinejad S, Bekhit AE-D, Greiner R, Mallikarjunan K. Potential application of pectin for the stabilization of nanoemulsions. Current Opinion in Food Science. 2018;**19**:72-76

[49] Yang X, Nisar T, Hou Y, Gou X, Sun L, Guo Y. Pomegranate peel pectin can be used as an effective emulsifier. Food Hydrocolloids. 2018;**85**:30-38

[50] Pereira PHF, Oliveira TÍS, Rosa MF, Cavalcante FL, Moates GK, Wellner N, et al. Pectin extraction from pomegranate peels with citric acid. International Journal of Biological Macromolecules. 2016;**88**:373-379

[51] Grassino AN, Barba FJ, Brnčić M, Lorenzo JM, Lucini L, Brnčić SR. Analytical tools used for the identification and quantification of pectin extracted from plant food matrices, wastes and by-products: A review. Food Chemistry. 2018;**266**:47-55

[52] Guolin H, Shi J, Kai Z, Xiaolan H. Application of ionic liquids in the microwave-assisted extraction of pectin from lemon peels. Journal of Analytical Methods in Chemistry. 2012;**2012**:302059

[53] Chan S-Y, Choo W-S. Effect of extraction conditions on the yield and chemical properties of pectin from cocoa husks. Food Chemistry. 2013;**141**(4):3752-3758

[54] Liu L, Jiang T, Yao J. A two-step chemical process for the extraction of cellulose fiber and pectin from Mulberry branch Bark efficiently. Journal of Polymers and the Environment. 2011;**19**(3):568

[55] Gopi D, Kanimozhi K, Bhuvaneshwari N, Indira J, Kavitha L. Novel banana peel pectin mediated green route for the synthesis of hydroxyapatite nanoparticles and their spectral characterization. Spectrochimica Acta Part A: Molecular and Biomolecular Spectroscopy. 2014;**118**:589-597

[56] Oliveira TÍS, Rosa MF, Cavalcante FL, Pereira PHF, Moates GK, Wellner N, et al. Optimization of pectin extraction from banana peels with citric acid by using response surface methodology. Food Chemistry. 2016;**198**:113-118

[57] Gharibzahedi SMT, Smith B, Guo Y. Pectin extraction from common fig skin by different methods: The physico-chemical, rheological, functional, and structural evaluations. International Journal of Biological Macromolecules. 2019;**136**:275-283

[58] Moorthy IG, Maran JP, Ilakya S, Anitha SL, Sabarima SP, Priya B. Ultrasound assisted extraction of pectin from waste Artocarpus heterophyllus fruit peel. Ultrasonics Sonochemistry. 2017;**34**:525-530

[59] Su D-L, Li P-J, Quek SY, Huang Z-Q, Yuan Y-J, Li G-Y, et al. Efficient extraction and characterization of pectin from orange peel by a combined surfactant and microwave assisted process. Food Chemistry. 2019;**286**:1-7

[60] Wicker L, Kim Y, Kim M-J, Thirkield B, Lin Z, Jung J. Pectin as a

bioactive polysaccharide – Extracting tailored function from less. Food Hydrocolloids. 2014;**42**:251-259

[61] Dubey V, Mohan P, Dangi JS, Kesavan K. Brinzolamide loaded chitosan-pectin mucoadhesive nanocapsules for management of glaucoma: Formulation, characterization and pharmacodynamic study. International Journal of Biological Macromolecules. 2020;**152**:1224-1232

[62] Villanova JCO, Ayres E, Oréfice RL. Design, characterization and preliminary in vitro evaluation of a mucoadhesive polymer based on modified pectin and acrylic monomers with potential use as a pharmaceutical excipient. Carbohydrate Polymers. 2015;**121**:372-381

[63] Hosseini S, Parastouei K, Khodaiyan F. Simultaneous extraction optimization and characterization of pectin and phenolics from sour cherry pomace. International Journal of Biological Macromolecules. 2020;**158**: 911-921

[64] Seixas FL, Fukuda DL, Turbiani FRB, Garcia PS, Petkowicz CLO, Jagadevan S, et al. Extraction of pectin from passion fruit peel (Passiflora edulis f. flavicarpa) by microwave-induced heating. Food Hydrocolloids. 2014;**38**:186-192

[65] Pérez CD, MD D'N, Rizzo SA, Gerschenson LN, Descalzo AM, Rojas AM. High methoxyl pectin–methyl cellulose films with antioxidant activity at a functional food interface. Journal of Food Engineering. 2013;**115**(1):162-169

[66] Cho E-H, Jung H-T, Lee B-H, Kim H-S, Rhee J-K, Yoo S-H. Green process development for apple-peel pectin production by organic acid extraction. Carbohydrate Polymers. 2019;**204**:97-103

[67] Talekar S, Vijayraghavan R, Arora A, Patti AF. Greener production of low methoxyl pectin via recyclable enzymatic de-esterification using pectin methylesterase cross-linked enzyme aggregates captured from citrus peels. Food Hydrocolloids. 2020;**108**:105786

[68] Zhao S, Yang F, Liu Y, Sun D, Xiu Z, Ma X, et al. Study of chemical characteristics, gelation properties and biological application of calcium pectate prepared using apple or citrus pectin. International Journal of Biological Macromolecules. 2018;**109**:180-187

[69] Mallick SP, Singh BN, Rastogi A, Srivastava P. Design and evaluation of chitosan/poly(l-lactide)/pectin based composite scaffolds for cartilage tissue regeneration. International Journal of Biological Macromolecules. 2018;**112**: 909-920

[70] Mellinas C, Ramos M, Jiménez A, Garrigós MC. Recent trends in the use of pectin from agro-waste residues as a natural-based biopolymer for food packaging applications. Materials. 2020;**13**(3):673

[71] Gutierrez-Pacheco MM, Ortega-Ramirez LA, Cruz-Valenzuela MR, Silva-Espinoza BA, Gonzalez-Aguilar GA, Ayala-Zavala JF. Chapter 50 - Combinational Approaches for Antimicrobial Packaging: Pectin and Cinnamon Leaf Oil. Antimicrobial Food Packaging. San Diego: Academic Press; 2016. pp. 609-617

[72] Laurent M, Patrick B. Stabilization mechanism of acid dairy drinks (ADD) induced by pectin. Food Hydrocolloids. 2003;**17**:445-454

[73] Ralet M-C, Lerouge P, Quéméner B. Mass spectrometry for pectin structure analysis. Carbohydrate Research. 2009;**344**(14):1798-1807

[74] Perussello CA, Zhang Z, Marzocchella A, Tiwari BK. Valorization

of Apple Pomace by extraction of valuable compounds. Comprehensive Reviews in Food Science and Food Safety. 2017;**16**(5):776-796

[75] Meneguin A, Cury B, Evangelista R. Films from resistant starch-pectin dispersions intended for colonic drug delivery. Carbohydrate Polymers. 2014;**99**:140-149

[76] Dranca F, Oroian M. Extraction, purification and characterization of pectin from alternative sources with potential technological applications. Food Research International. 2018;**113**:327-350

[77] Veisi Z, Gallant ND, Alcantar NA, Toomey RG. Responsive coatings from naturally occurring pectin polysaccharides. Colloids and Surfaces B: Biointerfaces. 2019;**176**:387-393

[78] Neckebroeck B, Verkempinck SHE, Vaes G, Wouters K, Magnée J, Hendrickx ME, et al. Advanced insight into the emulsifying and emulsion stabilizing capacity of carrot pectin subdomains. Food Hydrocolloids. 2020;**102**:105594

[79] Einhorn-Stoll U, Kastner H, Fatouros A, Krähmer A, Kroh LW, Drusch S. Thermal degradation of citrus pectin in low-moisture environment – Investigation of backbone depolymerisation. Food Hydrocolloids. 2020;**107**:105937

[80] Liu Z, Guo X, Meng H. Added ferulic acid enhances the emulsifying properties of pectins from different sources. Food Hydrocolloids. 2020;**100**:105439

[81] Maftoonazad N, Ramaswamy HS. Effect of pectin-based coating on the kinetics of quality change associated with stored avocados. The Journal of Food Processing and Preservation. 2008;**32**(4):621-643

[82] Sucheta, Rai SK, Chaturvedi K, Yadav SK. Evaluation of structural integrity and functionality of commercial pectin based edible films incorporated with corn flour, beetroot, orange peel, muesli and rice flour. Food Hydrocolloids. 2019;**91**:127-135

[83] Meerasri J, Sothornvit R. Characterization of bioactive film from pectin incorporated with gamma-aminobutyric acid. International Journal of Biological Macromolecules. 2020;(147):1285-1293

[84] Tripathi S, Mehrotra GK, Dutta PK. Preparation and physicochemical evaluation of chitosan/poly(vinyl alcohol)/pectin ternary film for food-packaging applications. Carbohydrate Polymers. 2010;**79**(3):711-716

[85] Lorevice MV, Otoni CG, Moura MR, Mattoso LHC. Chitosan nanoparticles on the improvement of thermal, barrier, and mechanical properties of high- and low-methyl pectin films. Food Hydrocolloids. 2016;**52**:732-740

[86] Ngo TMP, Nguyen TH, Dang TMQ, Tran TX, Rachtanapun P. Characteristics and antimicrobial properties of active edible films based on pectin and nanochitosan. International Journal of Molecular Sciences. 2020;**21**(6):2224

[87] Ranganathan P, Mutharani B, Chen S-M, Sireesha P. Biocompatible chitosan-pectin polyelectrolyte complex for simultaneous electrochemical determination of metronidazole and metribuzin. Carbohydrate Polymers. 2019;**214**:317-327

[88] Tsai R-Y, Chen P-W, Kuo T-Y, Lin C-M, Wang D-M, Hsien T-Y, et al. Chitosan/pectin/gum Arabic polyelectrolyte complex: Process-dependent appearance, microstructure analysis and its application. Carbohydrate Polymers. 2014;**101**:752-759

[89] Chen P-H, Kuo T-Y, Kuo J-Y, Tseng Y-P, Wang D-M, Lai J-Y, et al. Novel chitosan–pectin composite membranes with enhanced strength, hydrophilicity and controllable disintegration. Carbohydrate Polymers. 2010;**82**(4):1236-1242

[90] Chetouani A, Follain N, Marais S, Rihouey C, Elkolli M, Bounekhel M, et al. Physicochemical properties and biological activities of novel blend films using oxidized pectin/chitosan. International Journal of Biological Macromolecules. 2017;**97**:348-356

[91] Baron RD, Pérez LL, Salcedo JM, Córdoba LP, Sobral PJA. Production and characterization of films based on blends of chitosan from blue crab (Callinectes sapidus) waste and pectin from Orange (Citrus sinensis Osbeck) peel. International Journal of Biological Macromolecules. 2017;**98**:676-683

Chapter 6

Utilization of Agro-Industrial Wastes as Edible Coating and Films for Food Packaging Materials

Urmila Choudhary, Basant Kumar Bhinchhar, Vinod Kumar Paswan, Sheela Kharkwal, Satya Prakash Yadav and Prity Singh

Abstract

Mostly, food packaging employs synthetic materials obtained from nonrenewable sources. These packaging materials are based on petrochemicals and cause substantial environmental problems by producing massive amounts of non-biodegradable solid wastes. Edible coatings and films are considered as the potential solution to these problems of non-biodegradable packaging solid wastes for maintaining food-environment interactions, retaining food quality, and extending shelf life. In addition, edible coatings and films offer prevention from microbial spoilage of packed foods by controlling moisture and gas barrier characteristics. Increasing environmental concerns and consumer demands for high-quality eco-friendly packaging have fueled the advancement of innovative packaging technologies, for instance, the development of biodegradable films from renewable agricultural and food processing industry wastes. Therefore, the current chapter presents the application of edible coatings and films as an alternative to conventional packaging, emphasizing the fundamental characterization that these biodegradable packaging should hold for specific applications such as food preservation and shelf life enhancement. The primary employed components (e.g., biopolymers, bioactive, and additives components), manufacturing processes (for edible films or coatings), and their application to specific foods have all been given special consideration in this chapter. Besides, a future vision for the use of edible films and coatings as quality indicators for perishable foods is presented.

Keywords: agro-industrial wastes, edible films, edible coatings, food packaging, biopolymers

1. Introduction

Majority of food packaging are made of synthetic materials derived from nonrenewable sources and based on petrochemicals, having advantages of being available in large quantities at a low cost and having excellent barrier and resistance properties. Nevertheless, these are causing serious environmental issues due to the production of large amounts of non-biodegradable solid wastes [1]. Apart from its primary function of containing food, delaying deterioration, and extending shelf life, packaging also plays a critical role in regulating food-environment interactions [2, 3]. Environmental concerns and consumer demands for high-quality eco-friendly products that are similar to those found in nature (natural products), has prompted the development of technologies for novel packaging materials, such as the production of biodegradable films from renewable polymers [1, 3, 4]. As a result, consumer demand for packaging materials has switched to safe and environment friendly biodegradable materials, particularly from renewable agriculture by-products and food processing industry wastes. Polysaccharides including starch, cellulose, sodium alginate, pectin, chitosan, and gums, as well as proteins like whey, soy, gluten, and gelatin, are among the most commonly used biopolymers in the manufacture of biodegradable films [5]. Because of their abundance in nature, biodegradability, and edibility, these natural biopolymers are widely employed. Casting, pressing, and extrusion, followed by blowing, are some of the procedures utilized in the production of these films [6]. Plant-derived bioactive substances, such as essential oils, vitamins, minerals, polyphenols, and carotenoids, are extensively distributed in nature in addition to biopolymers. Because of their biological nature, different parts of plants, such as leaves, flowers, seeds, and roots, can possibly be employed in the manufacture of environment friendly films with functional features [7]. Some bioactive substances have antioxidant and antibacterial properties [8–10]. Bioactive films with new and/or improved properties, such as antioxidant [8, 9] and antimicrobial [10] effects, innovative colors [11, 12], and customized barrier and mechanical properties, have been developed using biopolymers and natural bioactive compounds [8, 11, 13]. Some of the techniques used for their production include the use of inherently bioactive biopolymer-based materials [14, 15], as well as the direct or sprinkling incorporation of free or encapsulated bioactive compounds into the film-forming solutions [8, 12].

Some polymers generated from renewable agro-waste sources are edible and have played important roles in food throughout history as well as in the food, pharmaceutical, and other industries. Biopolymers can be used alone or in combination with other biopolymers to produce an edible coating or film material [16, 17]. In comparison to non-edible polymeric packaging, edible coatings and films offer a number of advantages. They can simplify food packaging and, even if they are not consumed with the packaged product, they can assist to environmental pollution reduction due to their biodegradable nature. Material fragmentation and subsequent mineralization are the mechanisms through which polymers degrade in a bioactive environment. Outer temperature and moisture, as well as the enzyme activity of microorganisms degrade polymer, resulting in fragmented polymer residues. These polymer fragments are only considered biodegradable if they are consumed as food and for energy by microorganisms and converted into carbon dioxide (CO_2), water (H_2O), and biomass under aerobic conditions and hydrocarbons, methane, and biomass under anaerobic conditions at the end of the degradation process [18].

2. Agro-waste based renewable sources used in development of edible coatings and films

Bio-packaging films are made of materials derived from renewable resources that degrade completely. These can be made directly by biological systems (for example, plants, animals, algae, and microbes) or by polymerizing bio-based monomers (e.g.,

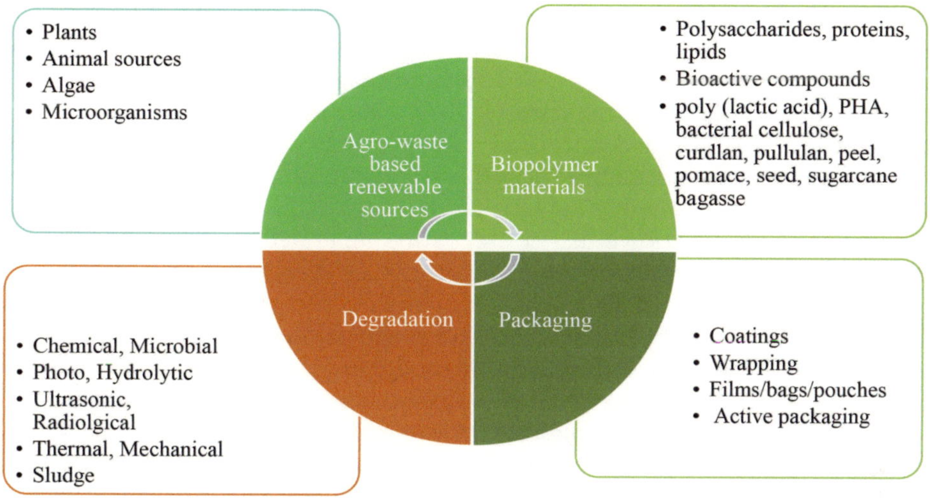

Figure 1.
Life cycle of agro-waste based renewable sources used in the development of edible coatings or films with modifications. (Adopted from Giosafatto et al. [19].)

polylactic acid). Classification and life cycle of various agro-waste based renewable sources used in the development of edible coatings or films has been depicted in **Figure 1**. These bio-polymer materials have been classified into four groups based on their origin and manufacturing method [20]. Among the biopolymers utilized in the packaging business include natural polysaccharides, proteins, and their derivatives [21, 22]. Polysaccharides (e.g., cellulose, pectin, gum, starch, chitosan) and proteins (e.g., collagen, casein, whey protein, egg protein, gluten) extracted from biomass, polymers synthesized from bio-derived monomers (e.g., polylactic acid (PLA)), and those produced directly by microorganisms (e.g., polyhydroxyalkanoates (PHA), pullulan, curdlan, bacterial cellulose) are examples of such materials based on renewable resources that are commonly used for food packaging applications [21]. These bio-based materials have good barrier properties and can be mass-produced on a large industrial scale for moderately low costs, making them a viable alternative to petroleum-based plastics. However, due to considerable divergence in respect to plastics, such as weak tensile strength, brittleness, thermal instability, and water sensitivity, commercialization of polymers produced from biomass is still limited [23–26]. As a result, various reinforcing materials and chemicals, such as plasticizers (e.g., glycerol, glycol, sorbitol), are combined with edible films and coatings to enhance their quality [27–31]).

In most circumstances, one of the above-mentioned biomaterials can be used alone or in combination with other biopolymers to create an edible coating or film material [16, 17].

3. Preparation of edible coatings and films

Edible coatings and films can be liquid, semi-solid or solid matrix that is wrapped around the surface of a food product and can be used as main packaging without providing any sensory or nutritional benefits. These are intended to be tasteless, colorless, non-toxic and unaffected by the sensory properties of the food product [32]. Trinetta [33] reported that when edible coatings and films are applied to cold or hot beverages, edible film dissolves and releases its contents, giving customers comfort portion control, and the elimination of solid waste. Recently, consumer awareness of edible, biodegradable, and environment friendly packaging materials has grown; edible films and coatings are increasingly being employed in the food processing industry for a wide range of food products. A comprehensive list of biopolymers used, their properties, functions and processing methods employed for production of edible coatings and films for various food products has been presented in **Table 1**. The coating method chosen has an impact not only on the preservation effect of the coating generated on the food products, but also on the cost of manufacturing and process efficiency. There are two types of edible packagings: (i) edible coatings applied directly to the food products and (ii) premade films wrapped around the food products. Two processes can be used to create edible coatings [54]. Dry-process methods, such as thermoplastic extrusion, rely on the thermoplastic characteristics of polymers when plasticized and heated above their glass-transition temperature in low-water-content circumstances. Extruded films have the drawback of not being able to cover uneven surfaces. On a commercial scale, extrusion and spraying are the predominant procedures for film creation and coating deposition, respectively. On the other hand, at a lab scale, the casting method for film creation and dipping methods for coating deposition, are simple to use and preferred methods.

Bio-polymers	Properties	Functions in edible coatings and films	Food product	Processing methods	References
Starch Cellulose Pectin Gums Chitosan Agar Alginate Dextran	Thickeners Gellants Stabilizers Coatings	They form base structure of a solid polymer matrix. They control physical changes, microbial growth, nutritional qualities and shelf-life.	Mango, Apple, Tomato, Strawberries, Green chilies, Sausage, Water melon, Plums, Bell pepper, Brinjal, Potatoes, Taro corms	Extrusion Solvent casting, co-extrusion, Dipping	[34–40]
Gelatin Pea protein isolate Casein Collagen Blood protein Fish protein Whey protein	Gellants Thickeners Stabilizers Emulsifiers Foaming	They help in transport of antimicrobials and antioxidants. They control transport of gases (mainly O_2)	Soybean oil Cheese slice, Sausage, Meat slice, Soluble coffee, Walnut kernels, Apple, Blood hake, Beef tenderloins, Pork loins, Salmon fillets Chicken breasts, Rice crisp balls	Solvent casting, Extrusion, Panning, Spraying, Dipping, Compression, Injection Molding, Electrospinning, 3D-printing	[5, 41–48]
Waxes Pullulan esters Chocolates Milk butter Oils (Olive oil and sunflower oil) Paraffin Glycerides	Protectors Coatings	They help to avoid drying or dehydration of the edible films and provide flexibility. They show anti-aging effects.	Strawberry, Fresh-cut apples	Solvent casting, Spreading, Dipping	[49, 50]

Bio-polymers	Properties	Functions in edible coatings and films	Food product	Processing methods	References
Additives (Glycerol, Resins, Polyglycerol Polyricinoleate, Lecithin, Ascorbic acid, Bioactive compounds)	Viscosity resistance Flexibility Antioxidants Stabilizers Anti-fungal Herbicides Fertilizers Antimicrobial Anti-inflammatory	They decrease the intermolecular force, melting temperature in the mixture and modify viscosity and rheological properties. They increase the solubility of polymers in water and work as stabilizers as well as protection for the products.	Pear, Mango, Quail eggs, Fresh- cut apples	Solvent casting, Spreading, Dipping	[49, 51–53]

Table 1.
List of biopolymers used, their properties, functions and processing methods employed for production of edible coatings and films for various food products.

4. Classification of edible coatings and films

Edible coatings and edible films are not the same; edible coatings can be applied directly to the surface of fruits, vegetables, and other food products in liquid form, usually by immersing the product in an edible material solution. Whilst edible films are molded as solid sheets and then utilized to wrap and package the food products [55]. The edible coating and film components are biodegradable and non-toxic. Biopolymer matrixes such as polysaccharides, proteins, lipids, and composite materials are used to create edible materials (**Table 1**). These coatings and films are thin layers created as a coating on a food surface or put (pre-made) between food components. Their goal is to enhance the shelf life of the food product while simultaneously acting as a safety barrier. They can retard moisture migration and the loss of volatile chemicals, as well as inhibit respiration and delay textural changes. In addition, as compared to typical synthetic films, they are good fat and oil barriers and have a high selective gas permeability ratio CO_2/O_2 [56]. They can also serve as carriers for food additives like antioxidants [57] and/or antimicrobial agents [10], as well as improve the product's mechanical integrity and handling properties. For some applications, stand-alone edible films with strong mechanical qualities could replace synthetic packaging films. Composition of biopolymers, their concentration, drainage time, viscosity, and other factors influence the mechanical and barrier properties of edible films and coatings [58].

5. Incorporation of bioactive compounds into edible coatings and films

Bioactive chemicals are generally secondary metabolites of plants that have both nutritional value and other functions in their metabolism, such as growth stimulant and protection against biotic and abiotic stress [59]. They are widely distributed in nature. Fruits and vegetables [11], leaves, blossoms [2], grains, seeds [60], rhizomes and roots [61], of different sorts of plants are important sources of bioactive components including phenols, proteins, essential oils, terpenoids and flavonoids. Plant-determined bioactive compounds are being viewed as fascinating elements for the creation of biodegradable and bioactive films because of their usefulness and natural origin [7]. Plant extracts and fruit pulps as sources of bioactive compounds or isolated bioactive compounds in film-forming solutions have been demonstrated to have antioxidant and antibacterial effects on the subsequent films, prolonging their utility in bioactive and biodegradable films or packaging [2, 3, 24, 62]. Plant derived naturally bioactive compounds are incorporated directly into agro-based polymers or the encapsulated plant-derived bioactive chemicals are integrated into the biopolymers by spraying during the production process [63]. Some edible coatings obtained from plant-derived bioactive compounds along with their methods of incorporation into the coatings, method of development of the edible coatings and films from these phytochemicals and the functional properties of these coatings and films are enlisted in **Table 2**.

Plant-derived bioactive compounds	Methods for incorporation	Methods of development of edible coatings and films	Functional properties	References
Blackberry pulp	Directly by sprinkling into the film-forming solution	Casting	Increased water vapor permeability and solubility in water; Reduced tensile strength; Antioxidant properties	[8, 12]
Curcumin	Directly into the film forming solution	Casting	Antioxidant and antimicrobial activity	[64]
Cranberry extract	Directly into the film forming solution	Casting	Antioxidant	[65]
Vitamin E (α-tocopherol encapsulated in carboxy methyl cellulose)	Directly into the film forming solution	Solution casting	Antioxidant; Decreased water vapor permeability and tensile strength	[66]
Tea Polyphenols	Directly into the film forming solution	Casting	Antioxidant and antimicrobial properties; Improved water barrier properties and tensile strength	[67, 68]
Babassu	Producing the film using inherently bioactive biopolymeric materials from agricultural by-products	Casting	Antioxidant	[15]
Essential oils	Directly into the film-forming solution	Casting	Improved barrier properties	[2]

Table 2.
List of some edible coatings obtained from plant-derived bioactive compounds, their methods of incorporation, method of development of the edible coatings and films and their functional properties.

6. Properties of edible coatings and films from agro-waste based polymers

In the food processing industry, edible coating/film provides a consistent quality for food items with market safety, nutritional value, and low manufacturing costs [69]. Control of mass transfers, mechanical protection, and sensory appeal are the most significant functions of an edible film or coating. Preventing desiccation of foods, regulating microenvironments of gases around foods, and controlling migration of ingredients and additives in food systems are all part of mass transfer control. Edible coatings on fresh foods can reduce quality changes and quantity losses by modifying and controlling the internal atmosphere of particular foods, which is an alternative to modified atmosphere storage. Even while penetration of oxygen may degrade food quality due to oxidation of the fragrance components in the food, modification of the internal atmosphere by the application of edible coatings can aggravate disorders associated with high carbon dioxide or low oxygen concentration. For fresh items, edible films with high water vapor permeability is also desired to extend the shelf life, while extremely high water vapor permeability is not, as it might cause significant moisture loss in the fruits during storage. The mechanical strength of an edible film must be sufficient to safeguard the packaging's integrity during distribution. The sensory qualities of an edible coating or film are critical aspects in final product approval.

The selection of ingredients is one of the most crucial aspects of making edible films. Polysaccharide-based edible coatings and films, a type of natural macromolecule with a high bioactivity, are often generated from agricultural feedstock or crustacean shell wastes. Polysaccharides that may form gels in water are found all over the plant kingdom. Some of them have been thoroughly explored, such as pectins in higher plants, carrageenans and agarose in algae, algal and bacterial alginates, and xanthans. Mucilages are heteropolysaccharides derived from plant stems, such as cactus stems. Food, cosmetics, pharmaceuticals, and other businesses may benefit from cactus mucilage [70]. The complex polysaccharide is a type of dietary fiber that can absorb adequate water before dissolving and dispersing and generating viscous or gelatinous colloids. The low cost of cactus mucilage as a coating is an important desirable consideration.

On the other hand, proteins such as casein, whey proteins, and maize zein, have been employed as a moisture barrier in edible coatings since they are numerous, abundant, inexpensive, and readily available. Corn zein, soy protein, wheat proteins (gluten, gliadin), peanut protein, gelatin, casein, and milk whey proteins have all been used in the production of protein-based films [5]. Food protein may act as natural vehicles, adapted to carry vital micronutrients (e.g. calcium and phosphate), building blocks (e.g. amino acids), as well as immune system components (e.g. immunoglobulins and lactoferrin) [71]. Furthermore, food proteins can be employed in coating formulations to create environment friendly packagings that are easily degradable and can be applied to a variety of foods such as vegetables, fruits, poultry, and fish items [72]. Protein-based edible films are appealing because, when compared to lipid- and polysaccharide-based edible films, they have superior gas barrier properties. Protein films' poor water vapor resistance and lower mechanical strength when compared to synthetic polymers, however, limit their use in food packaging.

Further, despite multifunctional potential of polysaccharides and proteins, the hydrophilic nature of these biopolymers limits their capacity to offer the desired edible film capabilities. Incorporation of hydrophobic chemicals, adjustment of polymer interaction, and production of cross-links are some of the ways to increase the water barrier and mechanical properties of these films. Lipid molecules used in edible coatings include neutral lipids of glycerides, which are esters of glycerol and fatty acids, and waxes, which are esters of long-chain monohydric alcohols and fatty acids. They are used to give hydrophobicity to food coatings [73]. Although protein films have good oxygen barrier and mechanical properties at low and intermediate relative humidity due to their large number of polar groups and extensive polymer inter-chain, the hydrophilic nature of proteins prevents edible protein films from acting as an effective water vapor barrier. However, lipid films have limited water vapor permeability due to their hydrophobic nature, but they are extremely brittle because of their monomeric structure. Furthermore, lipids generate opaque films or coatings and are susceptible to oxidation. These characteristics may affect the organoleptic characteristics of food and reduce their marketability. Natural antioxidants and antibacterial agents have also been added to the edible film to help prevent autoxidation of high-fat foods and boost oil resistance in fried foods [74, 75]. During storage, edible materials acts as barrier against moisture and gases from fresh produce, slowing enzymatic oxidation and protecting the food from browning and texture softening. These may also have the capacity to preserve natural volatile flavor compounds and prevent color components from discoloration [76]. Edible coatings and films aid to preserve phytochemical (phenolic, antioxidants, color) and physicochemical (total soluble solids, weight loss, pH, and respiration rate) attributes in fresh and minimally processed fruits and vegetables over time [77].

7. Physiochemical and morphological characterization of edible coatings and films

The main attributes and techniques for characterization of edible coatings and films are given below:

7.1 Physicochemical characterization

- Mechanical properties
- Solubility
- Color and transparency
- Thermal properties
- Microscopy
- Barrier properties

7.2 Performance evaluation

- Moisture loss, color, film thickness, microbial test
- Sensory properties
- Barrier properties (O_2 permeability, CO_2 permeability, water vapor permeability)

7.3 Morphological characterization

- Atomic force microscopy
- Transmission electron microscopy
- Scanning electron microscopy
- Universal tensile machine
- Fourier transform infrared spectroscopy
- Thermal methods
- Differential scanning calorimetry
- Thermogravimetric analysis
- Differential thermal analysis
- Dynamic mechanical analysis
- X-Ray diffraction
- Nuclear magnetic resonance spectroscopy analysis
- Dynamic light scattering
- Rheological analysis
- Zeta potential analysis

8. Application of edible coatings and films

Edible coatings and films produced from a variety of biopolymers can effectively preserve the nutritional and organoleptic qualities of various foods (**Figure 2**). Edible coatings and films are known as eco-friendly packaging materials, as they replace synthetic or plastic packaging materials and lower the post-harvest losses of fruits and vegetables [78–81]. The edible coating extends shelf life [76, 82–84], prevent microbiological contamination [85], minimizes lipid oxidation [86], and lowers their degradation effect [87].

Biopolymers-based edible coatings operate as barrier layers against gas diffusion, fragrance alterations, water migration, and various volatile exchange [88, 89]. Because of their great selective permeability to oxygen and carbon dioxide, polysaccharides have mostly been employed for food wrapping. The majority of these low-cost films are made from cellulose and its derivatives, such as ethers and esters, starch, pectins, and gums, which are used in food preservation. Fresh fruit products such as tomatoes, cherries, fresh beans, strawberries, mangoes, and bananas have all been coated with cellulose-based edible coatings to prevent quality loss. Chitosan is a polysaccharide that is commonly used to prevent post-harvest deterioration in fresh fruits and vegetables. Chitosan is made up of chitin, which is found in nature just next to cellulose in quantity [90]. Tahir et al. [91] investigated the efficiency of

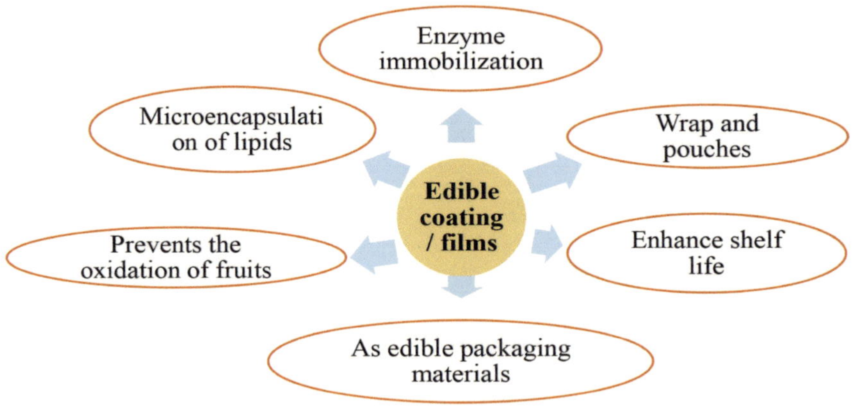

Figure 2.
Application of edible coating and films.

a gum arabic edible coating for increasing total antioxidant content in strawberry fruits during cold storage, with an increase in anthocyanin and phenolic contents. Furthermore, edible coatings containing chitosan and essential oils of oregano or thyme can inhibit the growth of spoilage and pathogen microorganisms while also improving the sensory quality of peeled shrimp [92]. Edible coatings are good transporter of functional ingredients and additives such color, flavor, minerals, vitamins, and antioxidant agents to improve the nutritional value, durability, and functionality of foods [77]. Ebrahimi and Rastegar [93] reported that a guar-based edible coating coated with *A. vera* and Spirulina platensis was effective in preserving the ascorbic acid content, total phenol, and antioxidant activity of mango maintained at room temperature. In addition, the coating extended the shelf life of mango fruit by reducing respiration and weight loss. Active films offer promising approach for slowly delivering the functional additives to the food surface, which could help to prevent food spoilage. Active packaging – a novel offshoot of the family of edible films - is quite encouraging as it can be carrier for a wide range of food additives, such as vitamins, antioxidants, minerals, colorants, fragrances, and antibacterial agents for the packaged food products [94].

9. Conclusions and future prospects

Starch, pectin, collagen, sodium alginate, gelatin, chitin, whey protein, chitosan, soy protein, gluten proteins and lipids are all examples of agro-industrial leftovers based biopolymers that have been widely used in the manufacturing of environment friendly food coatings and films. However, limited mechanical strengths and moisture barrier properties of most biopolymers-based edible coatings and films are the notable drawbacks. Combining agro- industrial leftovers based biopolymers with plant-derived bioactive substances (vitamins, carotenoids, phenolic compounds and phytochemicals, among others) permit the formation of bioactive films with antioxidants, antibacterial action. It is feasible to improve the mechanical and moisture barrier and physical properties of films by combining proteins (e.g., milk proteins, soy protein, collagen, and gelatin) with polysaccharides (e.g., starches, alginates, cellulose, and chitosan) or other polymers and hydrophobic compounds (lipids). Crosslinking procedures, on the other hand, could be a fascinating process that take into consideration chemical, enzymatic, and physical processes to produce biodegradable packaging materials with improved qualities from agro-industrial wastes.

Conflict of interest

The authors declare no conflict of interest.

Author details

Urmila Choudhary[1], Basant Kumar Bhinchhar[2], Vinod Kumar Paswan[3]*, Sheela Kharkwal[4], Satya Prakash Yadav[3] and Prity Singh[3]

1 Department of Food and Biotechnology, Faculty of Agriculture and Veterinary Science, Jayoti Vidyapeeth Women's University, Jaipur, India

2 Department of Livestock Production Management, Krishi Vigyan Kendra - Bharatpur, Sri Karan Narendra Agriculture University, Jobner, Jaipur, India

3 Department of Dairy Science and Food Technology, Institute of Agricultural Sciences, Banaras Hindu University, Varanasi, UP, India

4 Department of Agriculture Economics, Sri Karan Narendra Agriculture University, Jobner, Jaipur, India

*Address all correspondence to: vkpaswan.vet@gmail.com

References

[1] Zhao G, Lyu X, Lee J, Cui X, Chen WN. Biodegradable and transparent cellulose film prepared eco-friendly from durian rind for packaging application. Food Packaging and Shelf Life. 2019; 21:100345.

[2] Mahcene Z, Khelil A, Hasni S, Akman PK, Bozkurt F, Birech K, Goudjil MB, Tornuk F. Development and characterization of sodium alginate based active edible films incorporated with essential oils of some medicinal plants. International Journal of Biological Macromolecules. 2020; 145:124-132.

[3] Zahan KA, Azizul NM, Mustapha M, Tong WY, Rahman AMS, Sahuri IS. Application of bacterial cellulose film as a biodegradable and antimicrobial packaging material. Materials Today: Proceedings. 2020; 31:83-88.

[4] Gouveia TIA, Biernacki K, Castro MCR, Gonçalves MP, Souza HKS. A new approach to develop biodegradable films based on thermoplastic pectin. Food Hydrocolloids. 2019; 97:105175.

[5] Álvarez-Castillo E, Felix M, Bengoechea C, Guerrero A. (2021). Proteins from Agri-Food Industrial Biowastes or Co-Products and Their Applications as Green Materials. Foods. 2021; 10:981. DOI: 10.3390/foods10050981

[6] Fakhouri FM, Costa D, Yamashita F, Martelli SM, Jesus RC, Alganer K, Collares-Queiroz FP, Innocentini-Mei LH. (2013). Comparative study of processing methods for starch/gelatin films. Carbohydrate Polymers. 2013; 95:681-689.

[7] Jafarzadeh S, Jafari SM, Salehabadi A, Nafchi AM, Uthaya Kumar US, Khalil HPSA. (2020). Biodegradable green packaging with antimicrobial functions based on the bioactive compounds from tropical plants and their by-products. Trends in Food Science and Technology. 2020. 100:262-277.

[8] Nogueira GF, Soares CT, Cavasini R, Fakhouri FM, de Oliveira RA. Bioactive films of arrowroot starch and blackberry pulp: Physical, mechanical and barrier properties and stability to pH and sterilization. Food Chemistry. 2019(a); 275:417-425.

[9] Riaz A, Lagnika C, Luo H, Dai Z, Nie M, Hashim MM, Liu C, Song J, Li D. (2020) Chitosan-based biodegradable active food packaging film containing Chinese chive (*Allium tuberosum*) root extract for food application. International Journal of Biological Macromolecules. 2020; 150:595-604.

[10] Fu Y, Dudley EG. (2021). Antimicrobial-coated films as food packaging: A review. Comprehensive Review in Food Science and Food Safety. 2021; 1-34. DOI: 10.1111/1541-4337.12769

[11] Brito TB, Carrajola JF, Gonçalves ECBA, Martelli-Tosi M, Ferreira MSL. Fruit and vegetable residues flours with different granulometry range as raw material for pectin-enriched biodegradable film preparation. Food Research International. 2019; 121:412-421.

[12] Nogueira GF, Fakhouri FM, de Oliveira RA. Incorporation of spray dried and freeze dried blackberry particles in edible films: Morphology, stability to pH, sterilization and biodegradation. Food Packaging and Shelf Life. 2019(b); 20:100313.

[13] Zarandona I, Barba C, Guerrero P, de la Caba K, Maté J. Development of chitosan films containing-cyclodextrin

inclusion complex for controlled release of bioactives. Food Hydrocolloids. 2020; 104:105720

[14] Maniglia BC, Domingos JR, de Paula RL, Tapia-Blácido DR. (2014). Development of bioactive edible film from turmeric dye solvent extraction residue. LWT:Food Science and Technology. 2014; 56, 269-277.

[15] Maniglia BC, Tessaro L, Lucas AA, Tapia-Blácido DR. Bioactive films based on babassu mesocarp flour and starch. Food Hydrocolloids. 2017; 70:383-391.

[16] Shit SC, Shah PM. Edible polymers: Challenges and opportunities. Journal of Polymers. 2014; 427259:1-13.

[17] Pop OL, Pop CR, Dufrechou M, Vodnar DC, Socaci SA, Dulf FV, et al. Edible films and coatings functionalization by probiotic incorporation: A Review. Polymers. 2020; 12(12):1-15.

[18] Mohee R, Unmar GD, Mudhoo A, Khadoo P. Biodegradability of biodegradable/degradable plastic materials under aerobic and anaerobic conditions. Waste Management. 2008; 28:1624-1629.

[19] Giosafatto CVL, Antonio Fusco A, Al-Asmar A, Mariniello L. (2020). Microbial Transglutaminase as a Tool to Improve the Features of Hydrocolloid-Based Bioplastics. International Journal of Molecular Science. 2020; 21:3656. DOI:10.3390/ijms21103656

[20] DeGruson ML. Biobased polymer packaging. In Reference Module in Food Science; Smithers, G., Ed.; Elsevier Inc.: Amsterdam, The Netherlands. 2016.

[21] Gruji´c R, Vujadinovi´c D, Savanovi´c D. Biopolymers as food packaging materials. In Advances in Applications of Industrial Biomaterials; Pellicer E, Nikolic D, Sort J, Baró M, Zivic F, Grujovic N, Gruji´c R, Pelemis S. Eds.; Springer International Publishing: Cham, Switzerland, 2017; 139-160. ISBN 9783319627663.

[22] Pellicer E, Nikolic D, Sort J, Baró MD, Zivic F, Grujovic N, Gruji´c R, Pelemis S. Advances in Applications of Industrial Biomaterials, 1st ed.; Springer International Publishing: New York, NY, USA, 2017; ISBN 9783319627670.

[23] Jabeen N, Majid I, Nayik GA. (2015). Bioplastics and food packaging: A review. Cogent Food and Agriculture. 2015; 1:1117749.

[24] Chen H, Wang J, Cheng Y, Wang C, Liu H, Bian H, ..., Han W. (2019). Application of protein-based films and coatings for food packaging: A review. Polymers. 2019; 11(12):2039.

[25] Milani JM, Tirgarian B. (2020). An overview of edible protein-based packaging: Main sources, advantages, drawbacks, recent progressions and food applications. Journal of Packaging Technology and Research. 2020; 4:103-115.

[26] Zhong Y, Godwin P, Jin Y, Xiao H. Biodegradable polymers and green-based antimicrobial packaging materials: A mini-review. Advanced Industrial and Engineering Polymer Research. 2020; 3:27-35.

[27] Borges JA, Romani VP, Cortez-Vega WR, Martins VG. Influence of different starch sources and plasticizers on properties of biodegradable films. International Food Research Journal. 2015; 22(6).

[28] Jeyasubramanian K, Balachander R. Starch bioplastic film as an alternative food-packaging material. Journal of Achievements in Materials and Manufacturing Engineering. 2016; 75:78-84.

[29] Shendurse A. Milk protein based edible films and coatings—Preparation,

properties and food applications. Journal of Nutritional Health and Food Engineering. 2018; 8:219-226.

[30] Zakaria NH, Muhammad N, Sandu AV, Abdullah MMAB. Effect of mixing temperature on characteristics of thermoplastic potato starch film. IOP Conference Series: Materials Science and Engineering. 2018; 374:012083.

[31] Rahman R. Bioplastics for food packaging: A review. International Journal of Current Microbiology and Applied Sciences. 2019; 8:2311-2321.

[32] Viana RM, Sá NMSM, Barros MO, Borges MDF, Azeredo HMC. Nanofibrillated bacterial cellulose and pectin edible films added with fruit purees. Carbohydrate Polymers, 2018; 196:27-32.

[33] Trinetta V. Edible packaging. Reference Module in Food Sciences, 2016; 1-4.

[34] Aly SS, Mohamed EN, Abdou ES. Research Article Effect of Edible Coating on Extending the Shelf Life and Quality of Fresh Cut Taro. American Journal of Food Technology. 2017; 12:24-131.

[35] Calderón-Castro A, Vega-García MO, de Jesús Zazueta-Morales J, Fitch-Vargas PR, Carrillo-López A, Gutiérrez-Dorado R, ….., Aguilar-Palazuelos E. (2018). Effect of extrusion process on the functional properties of high amylose corn starch edible films and its application in mango (*Mangifera indica* L.) cv. Tommy Atkins. Journal of Food Science and Technology. 2018; 55(3): 905-914.

[36] Licciardello F, Lombardo S, Rizzo V, Pitino I, Pandino G, Strano MG, Muratore G, Restuccia C, Mauromicale G. Integrated agronomical and technological approach for the quality maintenance of ready-to-fry potato sticks during refrigerated storage. Postharvest Biology and Technology. 2018; 136:23-30.

[37] Fitch-Vargas, PR, Aguilar-Palazuelos E, Vega-García MO, Zazueta-Morales JJ, Calderón-Castro A, Montoya-Rodríguez A, et al. Effect of a corn starch coating obtained by the combination of extrusion process and casting technique on the postharvest quality of tomato. Revista Mexicana de Ingeniera Quimica, 2019. 18(3):789-801.

[38] Zhang W, Li X, Jiang W. Development of antioxidant chitosan film with banana peels extract and its application as coating in maintaining the storage quality of apple. International Journal of Biological Macromolecules. 2019.

[39] Hilbig J, Hartlieb K, Herrmann K, Weiss J, Gibis M. Influence of calcium on white efflorescence formation on dry fermented sausages with co-extruded alginate casings. Food Research International. 2020; 131(109012):1-8.

[40] Abral H, Pratama AB, Handayani D, Mahardika M, Aminah I, Sandrawati N, …., Ilyas RA. Antimicrobial Edible Film Prepared from Bacterial Cellulose Nanofibers/Starch/Chitosan for a Food Packaging Alternative. International Journal of Polymer Science, 2021. DOI: 10.1155/2021/6641284

[41] Antoniewski MN, Barringer SA, Knipe CL, Zerby HN. Effect of a gelatin coating on the shelf life of fresh meat. Journal of Food Science. 2007; 72(6):382-387.

[42] Chiu Po-en, Lih-shiuh L. Antimicrobial Activities of Tapioca Starch / Decolorized Hsian-Tsao Leaf Gum Coatings Containing Green Tea Extracts in Fruit-Based Salads, Romaine Hearts and Pork Slices. International Journal of Food Microbiology. 2010; 139:23-30.

[43] Carrión-Granda X, Fernández-Pan I, Rovira J, Maté J I.

(2018). Effect of antimicrobial edible coatings and modified atmosphere packaging on the microbiological quality of cold stored hake (merluccius merluccius) fillets. Journal of Food Quality2018; 1-12.

[44] Fadini AL, Rocha FS, Alvim ID, Sadahira MS, Queiroz MB, Alves RMV, et al. (2013). Mechanical Properties and Water Vapour Permeability of Hydrolysed Collagen-Cocoa Butter Edible Films Plasticised with Sucrose. Food Hydrocolloids. 30:625-631.

[45] Feng Z, Wu G, Liu C, Li D, Jiang B, Zhang X. Edible coating based on whey protein isolate nanofibrils for antioxidation and inhibition of product browning. Food Hydrocolloids, 2018; 79:179-188.

[46] Grosso, AL, Asensio CM, Nepote V, Grosso NR. Quality preservation of walnut kernels using edible coatings. Grasas Y Aceites. 2018; 69(4):1-9.

[47] Puscaselu R, Gutt G, Amariei S. Biopolymer-based films enriched with *Stevia rebaudiana* used for the development of edible and soluble packaging. Coatings. 2019; 9(6).

[48] Huntrakul K, Yoksan R, Sane A, Harnkarnsujarit N (2020). Effects of pea protein on properties of cassava starch edible films produced by blown-film extrusion for oil packaging. Food Packaging and Shelf Life. 2020; 24.

[49] Khan, MKI, Cakmak H, Tavman S, Schutyser M, Schroën K. Anti-browning and barrier properties of edible coatings prepared with electrospraying. Innovative Food Science and Emerging Technologies. 2014; 25:9-13.

[50] Niu B, Shao P, Chen H, Sun P. Structural and physiochemical characterization of novel hydrophobic packaging films based on pullulan derivatives for fruits preservation. Carbohydrate Polymers, 2019; 208:276-284.

[51] Dubey PK, Shukla RN, Srivastava G, Mishra AA, Pandey, A. Study on Quality Parameters and Storage Stability of Mango Coated with Developed Nanocomposite Edible Film. International Journal of Current Microbiology and Applied Sciences. 2019; 8(4):2899-2935.

[52] Yang K, Dang H, Liu L, Hu X, Li X, Ma Z, et al. (2019). Effect of syringic acid incorporation on the physical, mechanical, structural and antibacterial properties of chitosan film for quail eggs preservation. International Journal of Biological Macromolecules. 2019; 141:876-884.

[53] Rodríguez GM, Sibaja JC, Espitia PJP, Otoni CG. (2020). Antioxidant active packaging based on papaya edible films incorporated with Moringa oleifera and ascorbic acid for food preservation. Food Hydrocolloids, 2020; 103

[54] Antonio GC, Alves DG, Azoubel PM, Murr FEX, Park KJ. Influence of osmotic dehydration and high temperature short time processes on dried sweet potato (*Ipomoea batatas Lam.*). Journal of Food Engineering. 2008; 84(3):375-382.

[55] Aguirre-Joya JA, De Leon-Zapata MA, Alvarez-Perez OB, Torres-León C, Nieto-Oropeza DE, Ventura-Sobrevilla JM, ..., Aguilar CN. (2018). Basic and applied concepts of edible packaging for foods. In *Food packaging and preservation*. Academic Press. 2018; pp:1-61. DOI: 10.3390/foods10050981

[56] Falguera V, Quintero JP, Jimenez A, Munoz JA, Ibarz A (2011) Edible films and coatings: structures, active functions and trends in their use. Trends in Food Science and Technology 22(6): 292-303.

[57] Song Y, Liu L, Shen H, You J, Luo Y. Effect of sodium alginate-based edible

coatings containing different antioxidants on quality and shelf life of refrigerated bream (*Megalobrama amblycephala*). Food Control 2011; 22(3-4): 608-615.

[58] Lin D, Zhao Y. Innovations in the development and application of edible coatings for fresh and minimally processed fruits and vegetables. Comprehensive Reviews in Food Science and Food Safety, 2007; 6(3):60-75.

[59] Banožić M, Babić J, Jokić S. (2020). Recent advances in extraction of bioactive compounds from tobacco industrial waste-a review. Industrial Crops and Products. 2020; 144:112009.

[60] Rodsamran P, Sothornvit R. (2018). Bioactive coconut protein concentrate films incorporated with antioxidant extract of mature coconut water. Food Hydrocolloids. 2018; 79:243-252.

[61] Schaefer EW, Pavoni JMF, Luchese CL, Faccin DJL, Tessaro IC. Influence of turmeric incorporation on physicochemical, antimicrobial and mechanical properties of the cornstarch and chitosan films. International Journal of Biological Macromolecules. 2020; 148:342-350.

[62] Shirahigue LD, Ceccato-Antonini SR. Agro-industrial wastes as sources of bioactive compounds for food and fermentation industries. Ciência Rural. 2020; 50.

[63] Nogueira GF, de Oliveira RA, Velasco JI, Fakhouri FM. Methods of Incorporating Plant-Derived Bioactive Compounds into Films Made with Agro-Based Polymers for Application as Food Packaging: A Brief Review. Polymers. 2020; 12:2518. DOI:10.3390/polym12112518

[64] Nieto-Suaza L, Acevedo-Guevara L, Sánchez LT, Pinzón MI, Villa CC. (2019). Characterization of Aloe vera-banana starch composite films reinforced with curcumin-loaded starch nanoparticles. Food Structure. 2019; 22:100131.

[65] Wang S, Xia P, Wang S, Liang J, Sun Y, Yue P, Gao X. Packaging films formulated with gelatin and anthocyanins nanocomplexes: Physical properties, antioxidant activity and its application for olive oil protection. Food Hydrocolloids. 2019; 96:617-624.

[66] Mirzaei-Mohkam A, Garavand F, Dehnad D, Keramat J, Nasirpour A. Physical, mechanical, thermal and structural characteristics of nanoencapsulated vitamin E loaded carboxymethyl cellulose films. Progress in Organic Coatings. 2020; 138:105383.

[67] Farhan A, Hani NM. Active edible films based on semi-refined κ-carrageenan: Antioxidant and color properties and application in chicken breast packaging. Food Packaging and Shelf Life. 2020; 24:100476.

[68] Wu H, Lei Y, Zhu R, Zhao M, Lu J, Xiao D, Jiao C, Zhang Z, Shen G, Li S. Preparation and characterization of bioactive edible packaging films based on pomelo peel flours incorporating tea polyphenol. Food Hydrocolloids. 2019; 90:41-49.

[69] Bhardwaj A, Alam T, Talwar N. Recent advances in active packaging of agrifood products: A review. Journal of Postharvest Technology, 2019; 7(1):33-62.

[70] Del-Valle V, Hernández-Muñoz P, Guarda A, Galotto MJ. Galotto Development of a cactus-mucilage edible coating (*Opuntia ficus indica*) and its application to extend strawberry (*Fragaria ananassa*) shelf-life. Food Chemistry. 2005; 91:751-756.

[71] Livney YD. Milk proteins as vehicles for bioactives. Current Opinion in Colloid and Interface Science. 2010; 15(1-2):73-83.

[72] Rodriguez-Turienzo L, Cobos A, Moreno V, Caride A, Vieites JM, et al. Whey protein-based coatings on frozen Atlantic salmon (Salmo salar): Influence of the plasticiser and the moment of coating on quality preservation. Food Chemistry. 2011; 128(1):187-194.

[73] Aydin F, Kahve HI, Ardic M. Lipid Based Edible Films. Journal of Scientific and Engineering Research. 2017; 4(9):86-92.

[74] Daraei GA, Mirzaei HO, Maghsudlo Y, Kashaninejad M, Jafari SM. Production of low fat french-fries with single and multi-layer hydrocolloid coatings. Journal of Food Science and Technology. 2012; 1(8):660-669.

[75] Ganiari S, Choulitoudi E, Oreopoulou V. Edible and active films and coatings as carriers of natural antioxidants for lipid food. Trends in Food Science and Technology, 2017; 68:70-82.

[76] Sapper M, Amparo C. Starch-based coatings for preservation of fruits and vegetables. Coatings. 2018; 8(152):2-20.

[77] Sharma P, Shehin VP, Kaur N, Vyas P. Application of edible coatings on fresh and minimally processed vegetables: A review. International Journal of Vegetable Science, 2019; 25(3):295-314.

[78] Campos-Requena VH, Rivas BL, Pérez MA, Garrido-Miranda KA, Pereira ED. Polymer/clay nanocomposite films as active packaging material: Modeling of antimicrobial release. European Polymer Journal. 2015; 71:461-475.

[79] Medina JC, Gutiérrez TJ, Goyanes S, Bernal C, Famá L. Biodegradability and plasticizing effect of yerba mate extract on cassava starch edible films. Carbohydrate Polymers. 2016; 151:150-159.

[80] Bashir K, Jan K, Aggarwal M. Thermo-rheological and functional properties of gamma-irradiated whole wheat flour. International Journal of Food Science and Technology. 2017; 52(4):927-935.

[81] Moghadam M, Salami M, Mohammadian M, Khodadadi M, Emam-Djomeh Z. Development of antioxidant edible films based on mung bean protein enriched with pomegranate peel. Food Hydrocolloids. 2020. 105735.

[82] Grosso AL, Asensio CM, Grosso NR, Nepote V. Increase of walnuts' shelf life using a walnut flour protein-based edible coating. LWT:Food Science and Technology. 2020; 118.

[83] Hasan KSM, Ferrentino G, Scampicchio M. Nanoemulsion as advanced edible coatings to preserve the quality of fresh-cut fruits and vegetables: A review. International Journal of Food Science and Technology. 2020; 55:1-10.

[84] De Pilli T. Development of a vegetable oil and egg proteins edible film to replace preservatives and primary packaging of sweet baked goods. Food Control. 2020; 114.

[85] Bajpai S, Navin-Chand, Lodhi, R. Water sorption properties and antimicrobial action of zinc oxide nano particles loaded sago starch film. Journal of Microbiology, Biotechnology and Food Sciences. 2020; 2(4):2368-2387.

[86] Kazemian-Bazkiaee F, Ebrahimi A, Hosseini SM, Shojaee-Aliabadi S, Farhoodi M, Rahmatzadeh B, et al. Evaluating the protective effect of edible coatings on lipid oxidation, fatty acid composition, aflatoxins levels of roasted peanut kernels. Journal of Food Measurement and Characterization. 2020; 14(2):1025-1038.

[87] Dong C, Wang B, Li F, Zhong Q, Xia X, Kong B. Effects of edible chitosan

coating on Harbin red sausage storage stability at room temperature. Meat Science. 2020; 2020159.

[88] Sabbah M, Di Pierro P, Dell'Olmo E, Arciello, A, Porta R. Improved shelf-life of Nabulsi cheese wrapped with hydrocolloid films. Food Hydrocolloids. 2019; 96:29-35.

[89] Kõrge K, Bajići1 M, Likozar B, Novak U. Active chitosan-chestnut extract films used for packaging and storage of fresh pasta. 2020. DOI: 10.1111/IJFS.14569.

[90] Harish, KV, Prashanth, Tharanathan RN. Chitin/chitosan: modifications and their unlimited application potential-an overview. Trends in Food Science and Technology. 2007; 18(3):117-131.

[91] Tahir HE, Xiaobo Z, Jiyong S, Mahunu GK, Zhai X, Mariod AA. Quality and postharvest-shelf life of cold-stored strawberry fruit as affected by gum arabic (Acacia senegal) edible coating. Journal of Food Biochemistry. 2018; 42(3):12527.

[92] Carrión-Granda X, Fernández-Pan I, Jaime I, Rovira J, Maté JI. Improvement of the microbiological quality of ready-to-eat peeled shrimps (*Penaeus vannamei*) by the use of chitosan coatings. International Journal of Food Microbiology, 2016; 232:144-149.

[93] Ebrahimi F, Rastegar S. Preservation of mango fruit with guar-based edible coatings enriched with Spirulina platensis and Aloe-vera extract during storage at ambient temperature. Scientia Horticulturae. 2020; 265.

[94] Yong H, Liu J. (2021). Active packaging films and edible coatings based on polyphenol-rich propolis extract: A review. Comprehensive Review in Food Science and Food Safety. 2021; 20:2106-2145.

Section 4

Computational Applications

Chapter 7

Computational Applications for the Evaluation and Simulation of the Thermal Treatment of Canned Foods

William Miranda-Zamora, Amirpasha Tirado-Kulieva and David Ricse

Abstract

Throughout this chapter we will explore the computational applications that can help us in the evaluation, calculation and simulation of the thermal treatment of canned foods. Although some basic principles of microbial death kinetics will be recalled, the course is basically focused on the exploration and use of computational applications to evaluate and simulate the heat treatment of low-acid foods, considering *C. botulinum* as the reference microorganism. I hope that this book chapter will be useful for you and that you will be able to explore all the contents that are planned to be developed: General and technical aspects of the heat treatment of canned foods, heat penetration studies of canned foods, heat treatment evaluation General method, calculation and prediction of heat treatment by Ball's Method, heat treatment modeling and simulation, and optimization of heat treatment.

Keywords: canned food, heat penetration study, general method, Ball's Formula Method, simulation, F-value

1. Introduction

Heat treatment is a process of utmost importance to ensure food safety. If the product, a canned food, for example, does not receive an adequate heat treatment, it might cause intoxication and even death of the consumer [1]. The treatment must be designed correctly to guarantee efficient results, reducing the negative impact on the food, caused by the use of high temperatures [2]. For this, in order to optimize the process, it is necessary to know the thermal properties of the food, the kinetics of the changes in its quality, in addition to the quantitative and qualitative characteristics of the microbial load and/or enzymes [3].

To evaluate and simulate a thermal treatment, there are computational tools that use techniques that have been designed to evaluate the time of a process, or its F-value and/or simulate a thermal process.

In order to evaluate a thermal process there are two groups of methods, the Formula Methods and the General Methods. The **Figure 1** shows that there are several Formula methods to evaluate a heat treatment. The classic Formula Methods

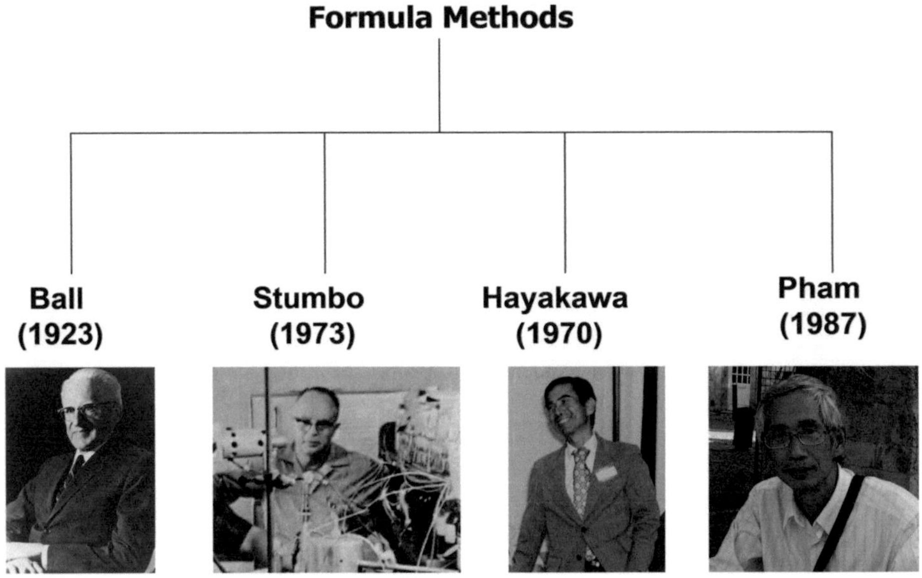

Figure 1.
Various formula methods to evaluate a thermal process of packaged foods.

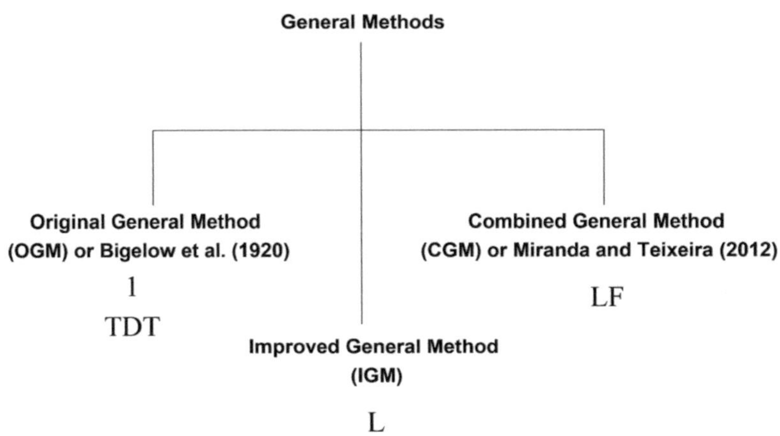

Figure 2.
Various general methods to evaluate a thermal process of packaged foods. TDT = thermal death time, L = lethal rate, and LF = LF-value.

are those of Charles Olin Ball and Charles Raymond Stumbo, which are called Ball's Formula Method [4–6], and Stumbo's Formula Method [7–10], respectively. There are other Formula Methods such as the one proposed by Kan-Ichi Hayakawa and the Formula Method developed by Quang Tuan Pham. These Formula Methods are called Hayakawa's Formula Method [11–13], and Pham's Formula Method [9, 13–15].

General methods can also be used to evaluate a packaged food. There are three General Methods (**Figure 2**). The Original General Method (OGM) was plated by Willard Dell Bigelow and his team in 1920 [15, 16]. The Improved General Method (IGM) was proposed by Charles Olin Ball after 1923 [15, 17–20].

The Combined General Method (CGM) was proposed by William R. Miranda-Zamora, and Arthur A. Teixeira in 2012, as its name says it is a combination of the two previous General Methods [2, 15].

Both General Methods and Formula Methods can be solved using software or computer applications [21–23].

The simulation of the heat treatment of packaged foods can be done using the finite difference method or using the finite element method [24–29].

2. General and technical aspects of the heat treatment of canned foods

Knowing the acidity of the food is crucial to determine the severity of the heat treatment. If the food has a pH < 4.6, as in the case of apple nectar, canned mango, pickles, citrus fruit juices, jam, sauerkraut, some sauces, etc.; in this case, the acidic medium will inhibit the proliferation of sporulated bacteria such as *C. botulinum* and, therefore, severe heat treatment ($<100°C$) will not be required. In contrast, if the food has low acidity (pH > 4.6), such as canned asparagus, milk, among other meat products, seafood and canned vegetables, it is essential to use high temperatures $>100°C$ [8, 14], which guarantee the sterilization of the food, increasing its shelf life. There are particular cases such as, for example, although milk has an almost neutral pH, being susceptible to microbiological spoilage, a mild pasteurization is applied to preserve its nutritional characteristics, in addition to the fact that milk is a product with a short shelf life; however, in many cases a high temperature-short time (HTST) pasteurization is applied and, being for a few seconds, the impact on the quality of the product is considerably avoided [2, 15].

Thermal treatment is based on two aspects: the biological (or microbiological) and the physical. The biological aspect refers to the microorganism, as the reference microorganism or target microorganism. The important thing is the number of decimal log reductions, n, [30–33], which is defined as:

$$n = \log\left(\frac{N_0}{N}\right) \tag{1}$$

where N_0 = the number of spores of the target microorganism, and N = the number of spores of the target microorganism that remains after heat treatment.

For example, if we start with a load of 1000 spores of *C. sporogenes*, and after heat treatment it is reduced to 1 spore. Therefore, the number of decimal log reductions will be 3 according to Eq. (1). In order to destroy the spores of *C. botulinum*, 12 decimal log reductions are needed [2, 15, 34–36]. The F_0 value is the standard used worldwide to quantify the F-value at a reference temperature. The reference temperature used for commercial sterilization is 121.1°C on the Celsius scale, or 250°F on the Fahrenheit scale. The F_0 value is the minutes that are necessary to evaluate the lethal effect of heat at the reference temperature of 121.1°C = 250°F [2, 15, 16, 21].

Other important values to determine are the D value and the z value, which are determined in the laboratory, using different methods. The D-value and z-value are derived from the thermal death or destruction curves and the thermal resistance curve, respectively [2]. The D-value is defined as the time required to destroy 90% of the initial microbial load, or go through a logarithmic cycle. The smaller the D value the faster the destruction rate [15]. The z-value is defined as the variation in degrees Celsius or degrees Fahrenheit required to reduce 90% of the D-value, or for the D-value to go through a logarithmic cycle [2, 15, 16].

One of the methods to determine the z-value and the decimal reduction time or D-value is the thermoresistometer [37–39]. The D-value and z-value can also be determined to the nutrients by means of a thermoresistometer [40–42]. The D value or decimal reduction time, is related to the number decimal log reductions through Eq. (2). Furthermore, the D-value and the z-value are characteristic of each spore or vegetative cell of the microorganism, or nutrient [43–45].

$$F_T^z = n \cdot \log\left(\frac{N_0}{N}\right) = n \cdot D \tag{2}$$

where F_T^z = F-value that depends on the temperature and the z-value of the microorganism. Generally, for low acid canned preserves a D value of 0.21 minutes and 12 decimal log reductions are taken, which gives a value of 2.52 minutes using Eq. (2). For handling in a food safety and process plant, it takes about F_T^z = 3 minutes [2, 15]. The physical aspect has to do with recording the temperature history using temperature sensors [46–50].

Cans and/or packaging have an important role in heat transfer efficiency and, therefore, they should be properly selected, considering their physical, mechanical, thermal and even electrical and optical characteristics. According to Berk [31], the most commonly used material is tinplate, especially because of its low cost. Aluminum is also widely used in the manufacture of cans for alcoholic and non-alcoholic beverages, and although it is more ductile and lighter than tinplate, it has a higher cost [18]. Glass is also used for the packaging of beverages and canned food, being characterized by its impermeability, rigidity, thermal resistance and transparency, obtaining attractive containers; however, they are very fragile and have a high weight [29]. Considering its thermal properties at 20°C, it has a thermal conductivity (k) of 0.75 W m^{-1} K^{-1}, a specific heat (C_p) of 800 J kg^{-1} K and a thermal diffusivity (α) of 0.35 x 106 m^2 s^{-1}, with a great difference compared to aluminum, whose k, C_p and α values are 230 W m^{-1} K^{-1}, 900 J kg^{-1} K and 95 x 106 m^2 s^{-1}, respectively. It is important to mention that these values depend on other characteristics, such as thickness, which, the higher the thickness, the better the thermal resistance, with a lower heat transfer rate [2, 16]. To avoid environmental impact, thinner materials are used, as in the case of tinplate, but like other metallic materials, they still maintain optimum qualities [34].

3. Heat penetration studies of canned foods

Heat penetration testing or studies is done by placing containers with temperature sensors in the coldest zone of the autoclave or retort. From the heat penetration tests it is interesting to determine the heat penetration factors f and j [51, 52].

The EVATMi-ZA v 1.0 software includes the determination of the heat penetration parameters f and j [1, 2] (heating, f_h and j_h [53–55] or cooling, f_c and j_c). The heat penetration parameters for heating include the determination of the delay factor $j_{\_CUT}$, based on the CUT value "come-up time" [56]. The come-up time is the time it takes for the autoclave or retort to reach process temperature. The 0.58 CUT is the new origin when using the Charles Olin Ball model or Ball's Formula Method [57–60]. **Figure 3** shows the behavior of steam within a canned food. The trend of temperature versus retort temperature or autoclave temperature is linear on semi logarithmic paper [61–64].

From **Figure 3** we deduce:

$$\frac{1}{f_h} = \frac{\log\left(j_h/j_{h\_CUT}\right)}{0.58CUT} \tag{3}$$

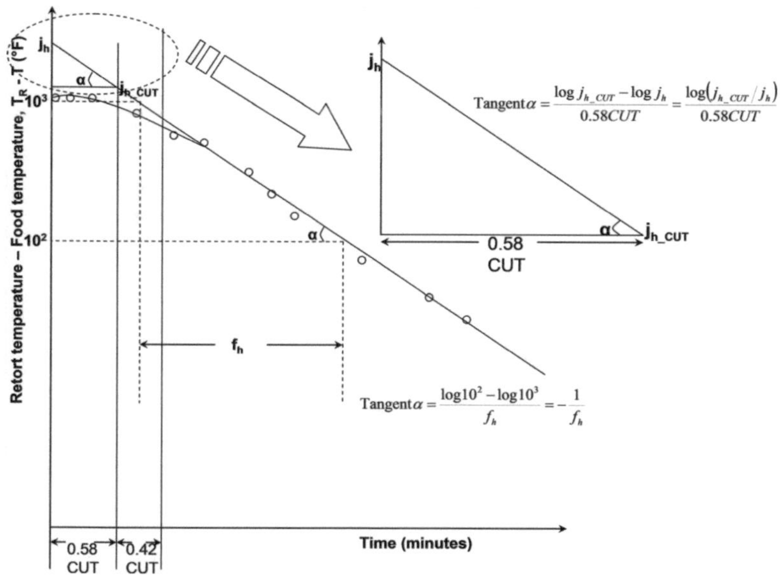

Figure 3.
Simple heating heat penetration curve.

Therefore, f_h as a function of CUT, j and $j_{\_CUT}$ is:

$$f_h = \frac{0.58CUT}{\log\left(j_h/j_{h\_CUT}\right)} \tag{4}$$

The EVATMi-ZA v 1.0 software includes the determination of the broken curve penetration parameters (f_h, f_{h2}, x_{bh}, x_{bh_CUT}, j_h, j_{h_CUT}). There are many canned foods that exhibit a broken curve [55, 65–70].

4. Heat treatment evaluation general method

Willard Dell Bigelow in 1920 presented a method for calculating the F-value of packaged foods, which was essentially graphical, it was determined by weighing (using scissors and analytical balance), counting squares (using graph paper) or planimetry (using planimeter) [71–74]. Initially, they constructed a graph in Cartesian coordinates, the curve of rise and fall of the temperature (heat penetration) at the slowest heating point (critical point) of the product during sterilization [75–80]. The thermal resistance of the bacteria was represented by the thermal destruction time curve (TDT-curve) obtained by plotting the time required to destroy a high percentage of spores from a population versus the degree temperature [81–83]. From the TDT-curve, the values of "thermal death or destruction time" (TDT) were calculated for each time of the heat penetration curve. This is known as the Original General Method (OGM).

To use the Original General Method, the Improved General Method and the Combined General Method it is necessary to calculate 1/TDT (min^{-1}), L (lethal rate) and LF (min) respectively. For which the following expressions will be used respectively [5, 15]:

$$\frac{1}{\text{TDT}} = \frac{10^{\frac{\text{T}-\text{T}_{\text{ref}}}{\text{z}}}}{\left(\text{F}^{\text{z}}_{\text{T}_{\text{ref}}}\right)_{Required}} \tag{5}$$

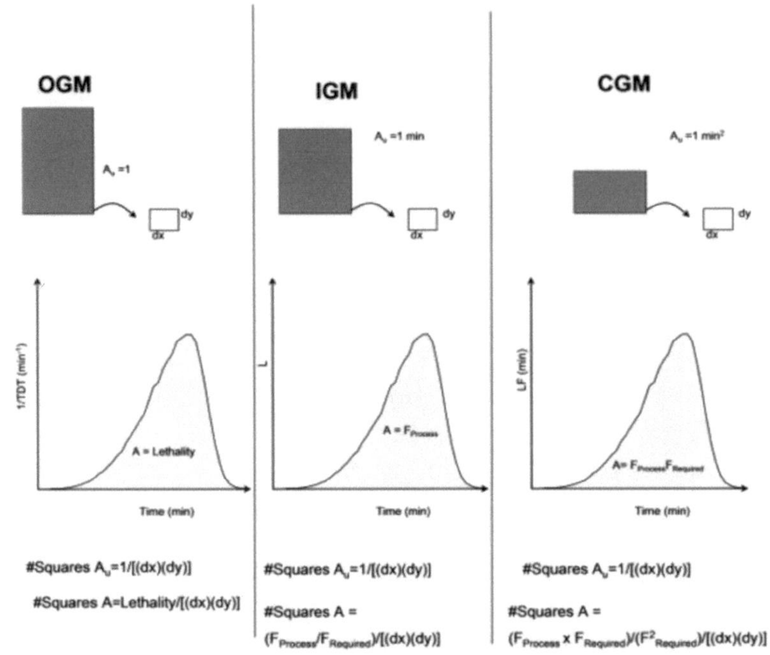

Figure 4.
General methods: Original general method (OGM), improved general method (IGM), and combined general method (CGM) using in a practical way the counting of squares on graph paper.

$$L = 10^{\frac{T - T_{ref}}{z}} \tag{6}$$

$$LF = \left(F^{z}_{T_{ref}}\right)_{Required} \times 10^{\frac{T - T_{ref}}{z}} \tag{7}$$

Figure 4 shows how to solve using General Methods using the counting of squares technique in a practical way. Also, it can be solved using numerical techniques such as Simpson's rule, the rectangular rule, or the trapezoidal rule. The EVATMi-ZA v 1.0 software can solve the General methods using numerical techniques (rectangular, trapezoidal and/or Simpson) [15].

5. Calculation and prediction of heat treatment by Ball's method

Charles Olin Ball developed and published his Formula Method in 1923 [84–90]. Olin Ball had already participated in research with Willard Dell Bigelow in 1920 [86]. The success of the Ball's Formula Method is that it appears in a Bulletin that goes directly to the canning industry [91]. Ball uses a hyperbola to represent the curvilinear part at the beginning of cooling. Ball does not use a hyperbolic function [91–93]. Ball's Formula Method is the favorite of the food industry. Ball's Formula Method is ninety-eight years old, it has crossed the threshold of time, although its handling is not well understood [91, 93]. Ball's Formula Method uses the heat penetration factors of heating. The Formula Method has two variants, simple curve and broken heating curve [94–96]. Formula Methods, such as Ball's Formula Method allow you to predict the F-value of the process, or calculate the process time. There are two cases to solve, or the time, or the F value. The Formula Methods are preferred to simulate or predict the process time, knowing the heat penetration parameters [97, 98]. The EVATMi-ZA v 1.0 software can solve time or F-value cases using Ball's Formula Method [99, 100].

6. Heat treatment modeling and simulation

Although heat treatments are of utmost importance due to their role in food preservation and also because they confer specific characteristics to food [101], the heat used during the process causes the degradation of nutrients, significantly affecting their quality; therefore, it is essential to minimize such damage, but without affecting the desired sterility of the products. For this purpose, it is necessary to develop a heat treatment that involves precise operating conditions in terms of temperature and time, achieving a minimal but efficient process [102], which guarantees optimum food quality [103]. Although currently it is still a challenge, thanks to technological advances, there are mathematical modeling and computer simulation techniques [104] that allow predicting the quality of food during processing and storage, in addition to providing the opportunity to optimize the process [105]. To achieve an ideal design, it is essential to have knowledge of the transport phenomena [106], such as mass, heat (by conduction, convection and radiation) and momentum, involved in the process [107], principles such as, for example, for solid foods, heat and moisture transfer, which are generally modeled by Fourier's and Fick's law, respectively [108] and for fluids, the continuity and Navier–Stokes Equations [109]. Likewise, the other physical mechanisms involved in the respective thermal process must be understood in order to achieve better results.

As it is known, preventing the deterioration of the organoleptic and nutritional characteristics of food [110], improves food safety, an important objective in the industry [108] and which is difficult to fulfill, due to the complex and dynamic nature of food [111] and thermal processing, requiring knowledge not only in engineering, but also in chemistry and microbiology [103]; therefore, as mentioned, numerical solutions are required [101] that, unlike traditional analytics, allow a fast and intelligent management of the large database obtained [112], helping to estimate the behavior of the food during thermal processing. It is necessary to mention that the pioneering study on the subject was by Datta & Teixeira [113], who performed a numerical modeling of the natural convection heating process of a liquid food packaged in cylindrical cans, successfully predicting the thermal (TP) and velocity (VP) profiles [114, 115].

By predicting and optimizing the process, modeling also helps to reduce time and costs, due to the reduction of experiments [116], which are very high under normal conditions [117], having as a disadvantage the obtaining of results in long periods [118]. Basically, one could, for example, after determining the effect of pasteurization temperature on the microbial load in different areas of a food, generate a mathematical model to help predict what the microbial spoilage would be if the processing temperature increases, without neglecting its influence on food quality. Likewise, a correct modeling enhanced with simulation, it would be possible to experiment with different changes in the variables, besides acquiring other advantages such as having control, and a broad and concise vision of the process [119].

For the execution of the computational techniques, first of all, the transfer phenomena that govern the thermal process, mainly heat, are represented by partial differential equations (PDE) [120] to be subsequently converted into a discrete model [107], which will be solved with some numerical method, such as finite differences, finite element and finite volume [108] or also called computational fluid dynamics (CFD) which is the most used [121], since it allows effectively designing a process or optimizing an existing one [122], through the development of three-dimensional models of the system and a numerical solution that describes it with high accuracy and realism. Similarly, it should be noted that, thanks to artificial intelligence, there are other modeling and optimization techniques, such as artificial neural networks and genetic algorithms, which are based on human intelligence and evolution, respectively [103].

CFD has been used since the 1950s, and has developed rapidly up to the present [117], since it is characterized by providing, through numerical algorithms, an easy resolution of the multiple physical phenomena involved in the process [120], which, in addition to heat, mass and momentum transfer (or fluid mechanics), also includes phase changes and chemical reactions [123]. Regarding thermal processing of canned foods, in which thermal processing is more difficult, because the temperature change is affected by the complex characteristics of the product [124], but also, the shape [125], type, size and orientation of the package [126]. CFD has been widely employed to solve challenges such as TP and VP monitoring [109], determination of the slowest heating zone (SHZ), slowest cooling zone (SCZ) [118] and even the kinetics of microbial inactivation and nutrient degradation. Specifically, in addition to sterilization [119] and pasteurization [127], it has covered a wide range of thermal treatments such as cooking [101] baking, drying [103] and cooling [128], and there is even information on its use in non-thermal processes such as microwave heating, ohmic heating, among others, which confirms its feasibility, versatility and suitability for food processing.

Since the since the development of CFD, there have been several commercial software, having in the early 1970s, a great boom and continuous improvement until today [127], emphasizing greater ease of use [129]. Of the extensive list of computer codes, some are FIDAP [116], CFX, FLUENT, PHOENICS [10], ANSYS, ANDINA-F, CFD++ [120], FLOW-3D, STAR-CD, CFD-ACE+ [128] and MSC Marc [105], of which most are still operational or have had some changes; for example, FLUENT and CFX are currently owned by ANSYS inc [129, 130], a leading developer of advanced engineering software. ANSYS offers several types of analysis and concerning heat treatments, it includes the three forms of heat transfer, phase change, internal heat source, contact thermal resistance, among other evaluations [131]. It is based on the creation of a geometry, which is divided into a finite number of units to form a computational geometry, then the governing PDEs are discretized, solved with numerical methods [122] and finally, the results are interpreted by the analyst. CFD simulation with ANSYS, has been applied to different packaged products such as in the pasteurization of beer [102] and water [115], in solid–liquid mixtures such as peas in water [132] and carrot-orange soup [133], food models such as waxy corn starch [126] and sucrose solution [134], in potato refrigeration [135], and even in the improvement of equipment processing parameters, such as in those of an industrial meat dryer [136] and a hydrofluidization freezing chamber [137]. There are also studies on the use of other software in solving the equations of energy, mass and moment, to determine the thermal behavior. A research deals with the effect of using 3.5% cornstarch packaging with an immobile can and rotating continuously at 146 rpm, using FIDAP 7.6 as software [138]. Furthermore, the same authors carried out a similar experiment, but evaluating the effect of intermittent axial agitation (0–146 rpm), and two retort temperatures (111 and 131°C) [139]. In another study, LabVIEW 8.5 was used to compare freezing results of guava pulp packed in stacked boxes, buckets, and unstacked drums, 34, 20, and 200 L, respectively [140].

7. Optimization of heat treatment

Until now, process optimization is essential to determine the best parameters that help to obtain the ideal results, in less time and with a significant reduction in costs. Considering the food industry, this task is much more difficult since there are multiple variables that intervene in the quality of the product [141] or process. Likewise, if it focuses mainly on the thermal processing of packaged foods and its importance in food safety, its optimization and especially its control are a challenge [142], due to the excessive use of high temperatures and for prolonged periods of time, and for Therefore, the improvement of the treatment conditions is essential to maintain the maximum characteristics of the food [143] and, consequently, a better acceptance by the consumer.

Regarding modeling and simulation, optimization, being related, for its application requires computational modeling and prediction techniques and that, due to advances in hardware, software and engineering related to processing thermal, it is becoming easier and faster to find the best solution [144]. In addition, one must have knowledge of heat transfer, quality change and microbial reduction, the three axes on which heat treatment is based.

Some techniques are the parametrization of the control vector, the principle of the continuous minimum, the super-simple optimization, the dynamic optimization, the neural network [145], the genetic algorithms, the simulated annealing of multiple initiation, among others. It should be noted that local optimization techniques have been used for thermal processing, which are the oldest and are hardly used, and global optimization techniques that are becoming increasingly popular [146]. This is due to the fact that, with the traditional ones, only the influence of a factor (independent variable) on a response (dependent variable) could be evaluated, therefore, as a general result of the processing was not obtained, the effect that the other variables had. This is problematic considering the complexity of the food, its dynamics [147] and also the characteristics of the container [145], including all the changes caused during heat treatment. For this, in contrast, global techniques [143], such as TOPSIS (Technique for Order of Preference by Similarity to Ideal Solution), focus on multiobjective optimization. In an investigation, it was applied in pressure pasteurization of green soybean tofu, managing to optimize the process with a thermal denaturation of 85°C and 10 min, and a high pressure homogenization of 80 MPa, for 4 cycles, achieving an increase in hardness, capacity of retention of water, proteins, fat and yield, in 155.7, 34.48, 30.31, 29.11 and 21.42%, respectively [148].

8. Conclusions

In this chapter we explore general and technical aspects of the heat treatment of canned foods, which become the basis or foundation. Heat penetration studies are the requirement for evaluation of heat treatment either by the General Method or by the Ball's Formula Method. Finally, in the final part, we review the modeling and simulation of the heat treatment, in order to achieve the optimization of the heat treatment. Thanks to the computational advances, it has been possible to improve the optimization techniques and even more the global ones, which are ideal to face the complexity of the thermal processing of packaged foods; however, there are still certain limitations such as the relative delay in executing the experimental design, due to the number of dependent and independent variables, with their respective levels (values) and replicas (repetitions).

Author details

William Miranda-Zamora*, Amirpasha Tirado-Kulieva and David Ricse
National University of Frontera, Sullana, Peru

*Address all correspondence to: wimiza23@hotmail.com

References

[1] Ling B, Tang J, Kong F, Mitcham EJ, Wang S. Kinetics of food quality changes during thermal processing: A review. Food and Bioprocess Technology. 2014; 8(2), 343–358. DOI: 10.1007/s11947-014-1398-3

[2] Tucker GS, Featherstone S, Miranda-Zamora WR. Fundamentos del procesamiento térmico de alimentos. Madrid: AMV; 2020. 516 p.

[3] van Boekel M, Fogliano V, Pellegrini N, Stanton C, Scholz G, Lalljie S, Somoza V, Knorr D, Jasti PR, Eisenbrand G. A review on the beneficial aspects of food processing. Molecular Nutrition & Food Research. 2010; 54(9), 1215–1247. DOI:10.1002/mnfr.200900608

[4] Stier RF. Technical and quality management of canning. Swainson's Handbook of Technical and Quality Management for the Food Manufacturing Sector. 2019; 505–527. DOI:10.1016/b978-1-78242-275-4.00020-4

[5] Pillai SD, Shayanfar S. Aseptic packaging of foods and its combination with electron beam processing. Electron Beam Pasteurization and Complementary Food Processing Technologies. 2015; 83–93. DOI:10.1533/9781782421085.2.83

[6] Hayakawa K, Ball CO. Charts for calculating average temperature of thermally conductive food in a cylindrical can during heat processing. Canadian Institute of Food Technology Journal. 1969; 2(1), 12–19. DOI:10.1016/s0008-3860(69)74336-7

[7] Sablani S., Shayya WH. Computerization of Stumbo's method of thermal process calculations using neural networks. Journal of Food Engineering. 2001; 47(3), 233–240. DOI: 10.1016/s0260-8774(00)00121-7

[8] Serment-Moreno V, Welti-Chanes J. Sterilization of foods. Encyclopedia of Food and Health. 2016; 175–180. DOI: 10.1016/b978-0-12-384947-2.00663-2

[9] Afaghi M, Ramaswamy, HS, Prasher SO. Thermal process calculations using artificial neural network models. Food Research International. 2001; 34(1), 55–65. DOI: 10.1016/s0963-9969(00)00132-0

[10] Mittal GS, Zhang J. Prediction of food thermal process evaluation parameters using neural networks. International Journal of Food Microbiology. 2002; 79(3), 153–159. DOI:10.1016/s0168-1605(02)00109-5

[11] Noronha J, Hendrickx M, Van Loey A, Tobback P. New semi-empirical approach to handle time-variable boundary conditions during sterilisation of non-conductive heating foods. Journal of Food Engineering. 1995; 24 (2), 249–268. DOI:10.1016/0260-8774(94)p2646-m

[12] Hayakawa K, Giannoni-Succar EB, Huang F, Zhou L. Use of the empirical temperature response function for modified Duhamel's theorem application. Journal of Food Engineering. 1997; 34(3), 331–353. DOI: 10.1016/s0260-8774(97)00084-8

[13] Rattan NS. Heating behavior and quality changes in canned potatoes subjected to agitation processing [thesis]. Montreal, Canada: McGill University; 2012.

[14] Stoforos NG. Thermal process design. Food Control. 1995; 6(2), 81–94. DOI:10.1016/0956-7135(95)98911-j

[15] Miranda-Zamora WR, Teixeira AA. Principios matemáticos del proceso térmico de alimentos. Madrid: AMV; 2012. 560 p.

[16] Heldman DR, Hartel RW. Principles of food processing. Gaithersburg, Maryland: Aspen Publishers, Inc; 1997. 288.

[17] Friso D. A new mathematical model for food thermal process prediction. Modelling and Simulation in Engineering. 2013; 1–8. DOI:10.1155/2013/569473

[18] Simpson R, Almonacid S, Teixeira A. Bigelow's General method revisited: Development of a New Calculation Technique. Journal of Food Science. 2003; 68(4), 1324–1333. DOI:10.1111/j.1365-2621.2003.tb09646.x

[19] Körmendy L, Zsarnóczay G, Mihályi V. A new, modified acid phosphatase assay for determining the extent of heat treatment in canned hams. Food Chemistry. 1992; 44(5), 367–375. DOI:10.1016/0308-8146(92)90270-c

[20] Simpson R, Figueroa I, Teixeira A. Optimum on-line correction of process deviations in batch retorts through simulation. Food Control. 2006; 17(8), 665–675. DOI:10.1016/j.foodcont.2005.06.004

[21] Miranda Zamora WR, Sanchez Chero MJ, Sanchez Chero JA. Software for the determination of the time and the F value in the thermal processing of packaged foods using the Modified Ball Method. In: Ahram T., Karwowski W., Vergnano A., Leali F., Taiar R. (eds) Intelligent Human Systems Integration 2020. IHSI 2020. Advances in Intelligent Systems and Computing. 2020; vol 1131. Springer, Cham. DOI:10.1007/978-3-030-39512-4_78

[22] Zamora WRM, Villarreyes SSC, Povis NLL, More LAV, Chero MJS, Panca CMA, Morales MVS. A new mathematical solution for packaged food thermal processing. In: Mrugalska B., Trzcielinski S., Karwowski W., Di Nicolantonio M., Rossi E. (eds) Advances in Manufacturing, Production Management and Process Control. AHFE 2020. Advances in Intelligent Systems and Computing. 2020; vol 1216. Springer, Cham. DOI:10.1007/978-3-030-51981-0_49

[23] Zamora WRM, Chero MJS, Timaná-Alvarez M, Seminario-Morales V, Niño-Carmona C, Leyva N, More LAV, Ticona-Carrizales L, Ygnacio A. Program in Visual Basic Language: A Simplified Procedure for Thermal Treatment Evaluation of Packaged Foods. In: Russo D., Ahram T., Karwowski W., Di Bucchianico G., Taiar R. (eds) Intelligent Human Systems Integration 2021. IHSI 2021. Advances in Intelligent Systems and Computing. 2021; vol 1322. Springer, Cham. DOI:10.1007/978-3-030-68017-6_71

[24] Özişik MN, Orlande HRB, Colaço MJ, Cotta RM. Finite difference methods in heat transfer. 2nd Edition. Boca Raton, FL: CRC Press; 2017. 600.

[25] Welt BA, Teixeira AA, Chau KV, Balaban MO, Hintenlang DE. Explicit finite difference methods for heat transfer simulation and thermal process design. Journal of Food Science. 1997; 62 (2), 230–236. DOI:10.1111/j.1365-2621.1997.tb03974.x

[26] Gosz MR. Finite Element Method. Boca Raton, FL: CRC Press; 2006. 400

[27] Banga JR, Alonso AA, Gallardo JM, Perez-Martin RI. Mathematical modelling and simulation of the thermal processing of anisotropic and non-homogeneous conduction-heated canned foods: Application to canned tuna. Journal of Food Engineering. 1993; 18(4), 369–387. DOI:10.1016/0260-8774(93)90053-m

[28] Tucker GS. Development and use of numerical techniques for improved thermal process calculations and

control. Food Control. 1991; 2(1), 15–19. DOI:10.1016/0956-7135(91)90113-b

[29] Teixeira AA, Dixon JR, Zabradnik JW, Zinsmeister GE. Computer optimization of nutrient retention in the thermal processing of conduction heated foods. Food Techology. 1969; 23, 137–142.

[30] Sekhon AS, Singh A, Michael M. Short communication: Decimal log reductions of Salmonella Senftenberg 775 W and other Salmonella serovars in nonfat milk and powder. Journal of Dairy Science. 2020; 103, 8, 6894–6899. DOI:10.3168/jds.2019-17844

[31] Berk Z. Thermal processing. Food Process Engineering and Technology. 2018; 399–420. DOI:10.1016/b978-0-12-812018-7.00017-8

[32] Sant'Ana AS, Rosenthal A, Massaguer PR. Heat resistance and the effects of continuous pasteurization on the inactivation of Byssochlamys fulva ascospores in clarified apple juice. Journal of Applied Microbiology. 2009; 107(1), 197–209. DOI:10.1111/j.1365-2672.2009.04195.x

[33] Rachon G. Survival of pathogens in low moisture foods [thesis]. Whiteknights, United Kingdom: University of Reading; 2017.

[34] Tucker G, Featherstone S. Essentials of thermal processing. Second edition. Chichester, UK: Wiley-Blackwell; 2021. 352.

[35] Diao MM, André S, Membré J-M. Meta-analysis of D-values of proteolytic *Clostridium botulinum* and its surrogate strain *Clostridium sporogenes* PA 3679. International Journal of Food Microbiology. 2014; 174, 23–30. DOI: 10.1016/j.ijfoodmicro.2013.12.029

[36] Rosnes JT, Fernandez PS, Periago PM, Shara T. Microorganisms of relevance in thermally processed foods. In: Valdramidis, V., Van Impe, J.F.M. (Eds.), Progress on Quantitative Approaches of Thermal Food Processing. Nova Science Publishers, New York; 2012. p. 1–37.

[37] Condón S, Arrizubieta MJ, Sala FJ. Microbial heat resistance determinations by the multipoint system with the thermoresistometer TR-SC Improvement of this methodology. Journal of Microbiological Methods. 1993; 18(4), 357–366. DOI:10.1016/0167-7012(93)90017-c

[38] André S, Leguerinel I, Palop A, Desriac N, Planchon S, Mafart P. Convergence of Bigelow and Arrhenius models over a wide range of heating temperatures. International Journal of Food Microbiology. 2019; 291, 173–180. DOI:10.1016/j.ijfoodmicro.2018.11.019

[39] Garre A, González-Tejedor GA, Aznar A, Fernández PS, Egea JA. Mathematical modelling of the stress resistance induced in Listeria monocytogenes during dynamic, mild heat treatments. Food Microbiology. 2019; 84, 103238. DOI:10.1016/j.fm.2019.06.002

[40] Al Fata N, Georgé S, André S, Renard CMGC. Determination of reaction orders for ascorbic acid degradation during sterilization using a new experimental device: The thermoresistometer Mastia ®. LWT - Food Science and Technology. 2017; 85, 487–492. DOI:10.1016/j.lwt.2016.08.043

[41] Blasco R, Esteve MJ, Frígola A, Rodrigo M. Ascorbic acid degradation kinetics in mushrooms in a high-temperature short-time process controlled by a thermoresistometer. LWT - Food Science and Technology. 2004; 37(2), 171–175. DOI:10.1016/j.lwt.2003.08.003

[42] Al Fata N, Georgé S, Dlalah N, Renard CMGC. Influence of partial pressure of oxygen on ascorbic acid

degradation at canning temperature. Innovative Food Science & Emerging Technologies. 2018; 49, 215–221. DOI: 10.1016/j.ifset.2017.11.007

[43] Gabriel AA, Ubana MA. Decimal reduction times of Salmonella Typhimurium in guinataang kuhol: An indigenous Filipino dish. LWT - Food Science and Technology. 2007; 40(6), 1108–1111. DOI:10.1016/j.lwt.2006.06.003

[44] Membré J-M, Diao M, Thorin C, Cordier G, Zuber F, André S. Risk assessment of proteolytic *Clostridium botulinum* in canned foie gras. International Journal of Food Microbiology. 2015; 210, 62–72. DOI: 10.1016/j.ijfoodmicro.2015.06.002

[45] Wei X, Lau SK, Chaves BD, Danao M-GC, Agarwal S, Subbiah J. Effect of water activity on the thermal inactivation kinetics of Salmonella in milk powders. Journal of Dairy Science. 2020; 103, 8, 6904–6917. DOI:10.3168/jds.2020-18298

[46] Mulla R, Dunnill CW. Single material thermocouples from graphite traces: Fabricating extremely simple and low cost thermal sensors. Carbon Trends. 2021; 4, 100077. DOI:10.1016/j.cartre.2021.100077

[47] Fryer PJ, Simmons MJH, Cox PW, Mehauden K, Hansriwijit S, Challou F, Bakalis S. Temperature Integrators as tools to validate thermal processes in food manufacturing. Procedia Food Science. 2011; 1, 1272–1277. DOI: 10.1016/j.profoo.2011.09.188

[48] Tarzan-Lorente M, Ceravalls J, Bosch J, Cama JMG, Pardo A. Electronic system for controlling the food cooking process. Procedia Chemistry. 2009; 1 (1), 489–492. DOI:10.1016/j.proche.2009.07.122

[49] Gil AG, Ochoa González OA, Cardona Sepúlveda LF, Alvarado Torres PN. Venting stage experimental study of food sterilization process in a vertical retort using temperature distribution tests and energy balances. Case Studies in Thermal Engineering. 2020; 100736. DOI:10.1016/j.csite.2020.100736

[50] Sullivan JJ. Wireless data loggers to study heat penetration in retorted foods. In-Pack Processed Foods. 2008; 116–130. DOI:10.1533/9781845694692.2.116

[51] Stumbo CR. Thermobacteriology in food processing. Second edition. New York: Academic Press; 1973. 329.

[52] Etzel MR, Willmore P, Ingham BH. Heat penetration and thermocouple location in home canning. Food Science & Nutrition. 2014; 3(1), 25–31. DOI: 10.1002/fsn3.185

[53] Ranganathan K, Rangaswamy S, Subramanian V, Shanmugam N. Modelling of drying kinetics and heat penetration studies on carrot. International Journal of Engineering and Technical Research. 2015; 3(5), 371–376.

[54] Awuah GB, Khurana A, Weddig LM, Balestrini CG. A comparative study of heat penetration data using remote sensors and needle or rod-in-tube thermocouples. Journal of Food Process Engineering. 2007; 30(4), 458–471. DOI:10.1111/j.1745-4530.2007.00106.x

[55] Smout C, Ávila I, Van Loey AML, Hendrickx, MEG, Silva C. Influence of rotational speed on the statistical variability of heat penetration parameters and on the non-uniformity of lethality in retort processing. Journal of Food Engineering. 2000; 45(2), 93–102. DOI:10.1016/s0260-8774(00)00045-5

[56] Simpson R, Almonacid S, Nuñez H, Urtubia A, Teixeira AA. Is there a need for the come-up time correction factor in Ball's Formula Method? A critical

analysis. Food Engineering Reviews. 2012; 4(2), 107–113. DOI:10.1007/s12393-012-9049-9

[57] Radrigan R. Computer simulation of thermal processing for food. Heat Transfer Phenomena and Applications. 2012; 183–202.DOI:10.5772/51815

[58] Dixon WR, Watts EG, King JA, Fu X, Wicker L. Shelf-stable sustainable shrimp thermally processed with reciprocal agitation. Frontiers in Sustainable Food Systems. 2020; 4, 1–12 DOI:10.3389/fsufs.2020.569790

[59] Adepoju MA, Omitoyin BO, Mohan CO, Zynudheen AA. Heat penetration attributes of milkfish (Chanos chanos) thermal processed in flexible pouches: a comparative study between steam application and water immersion. Food Science & Nutrition. 2016; 5(3), 521–524. DOI:10.1002/fsn3.426

[60] Berry MR. Prediction of Come-Up time correction factors for batch-type agitating and still retorts and the influence on thermal process calculations. Journal of Food Science. 1983; 48(4), 1293–1299. DOI:10.1111/j.1365-2621.1983.tb09214.x

[61] Datta AK. On the theoretical basis of the asymptotic semilogarithmic heat penetration curves used in food processing. Journal of Food Engineering. 1990; 12(3), 177–190. DOI:10.1016/0260-8774(90)90009-w

[62] Körmendy I, Körmendy L. Considerations for calculating heat inactivation processes when semilogarithmic thermal inactivation models are non-linear. Journal of Food Engineering. 1997; 34(1), 33–40. DOI:10.1016/s0260-8774(97)00071-x

[63] Körmendy I, Körmendy L, Ferenczy A. Thermal inactivation kinetics of mixed microbial populations. A hypothesis paper. Journal of Food Engineering. 1998; 38(4), 439–453. DOI:10.1016/s0260-8774(98)00119-8

[64] Downing DL. Heat penetration determinations and thermal process calculations. A Complete Course in Canning and Related Processes. 1996; 39–102. DOI:10.1533/9781845696207.39

[65] Reynaga W. Estudio del tratamiento térmico de enlatado de pechuga de pollo (*Gallus gallus*) en trozos y desmenuzado [thesis]. La Molina, Lima: Universidad Nacional Agraria La Molina; 2014.

[66] Yang WH, Rao MA. Numerical study of parameters affecting broken heating curve. Journal of Food Engineering. 1998; 37(1), 43–61. doi:10.1016/s0260-8774(98)00070-3

[67] Berry MR, Bush RC. Establishing thermal processes for products with broken-heating curves from data taken at other retort and initial temperatures. Journal of Food Science. 1987; 52(4), 958–961. DOI:10.1111/j.1365-2621.1987.tb14252.x

[68] Denys S, Noronha J, Stoforos NG, Hendrickx M, Tobback P. Evaluation of process deviations, consisting of drops in rotational speed, during thermal processing of foods in rotary water cascading retorts. Journal of Food Engineering. 1996; 30(3–4), 327–338. DOI:10.1016/s0260-8774(96)00057-x

[69] Wiese KL, Wiese, KF. A comparison of numerical techniques to calculate broken line heating factors of a thermal process. Journal of Food Processing and Preservation. 1992; 16(5), 301–312. DOI:10.1111/j.1745-4549.1992.tb00211.x

[70] Llave YA, Hagiwara T, Sakiyama T. Artificial neural network model for prediction of cold spot temperature in retort sterilization of starch-based foods. Journal of Food Engineering. 2012; 109(3), 553–560. DOI:10.1016/j.jfoodeng.2011.10.024

[71] Fellows PJ. Heat sterilisation. Food Processing Technology. 2017; 581–622. DOI:10.1016/b978-0-08-100522-4.00012-2

[72] Kramer A, Twigg BA. Principles and instrumentation for the physical measurement of food quality with special reference to fruit and vegetable products. Advances in Food Research. 1960; 9, 153–220. DOI:10.1016/s0065-2628(08)60276-1

[73] Stumbo CR. The General Method. Thermobacteriology in Food Processing. 1973; 143–151. DOI:10.1016/b978-0-12-675352-3.50019-x

[74] Biran A. Geometric properties of areas and volumes. Geometry for Naval Architects. 2019; 121–194. DOI:10.1016/b978-0-08-100328-2.00012-2

[75] Rinaldi M, Chiavaro E, Massini R. Real-time estimation of slowest heating point temperature and residual cooking time by coupling multipoint temperature measurement and mathematical modelling: Application to meat cooking automation. Food Control. 2012; 23(2), 412–418. DOI:10.1016/j.foodcont.2011.08.009

[76] Huang XJ, Hanzawa T, Sakai N. The characteristics of the slowest heating point in a canned food in oil. Nippon Shokuhin Kogyo Gakkaishi. 1992; 39(1), 1–7. DOI:10.3136/nskkk1962.39.1

[77] Hanzawa T, Wang QH, Suzuki M, Sakai N. Numerical analysis of slowest heating or cooling point in a canned food in oil. Journal of Chemical Engineering of Japan. 1998; 31(3), 451–455.DOI:10.1252/jccj.31.451

[78] Wang Q-Z, Hanzawa T, Sakai N. Estimating heating time for sterilizing process in a canned liquid foods with particles. Nippon Shokuhin Kagaku Kogaku Kaishi. 1998; 45(11), 676–682. DOI:10.3136/nskkk.45.676

[79] Assan MYA, Watanabe H, Mihori T. Temperature distribution at the surface of cans in an industrial scale static retort during saturated steam sterilization. Food Science and Technology Research. 2000; 6(3), 196–200. DOI:10.3136/fstr.6.196

[80] Yasui T, Esselen WB, Fukazawa T, Hashimoto Y. Processing studies on canned corned beef and canned luncheon meat. Agricultural and Biological Chemistry. 1961; 25(8), 632–636. DOI:10.1271/bbb1961.25.632

[81] Hayakawa K. Selective review of research results related to thermal process lethality estimation. Japan Journal of Food Engineering. 2001; 2(2), 47–52. DOI:10.11301/jsfe2000.2.47

[82] Yamamoto Y, Ono N, Higashi K, Yoshii H. Studies on growth inhibition of food spoilage microorganisms for low salt foods. Part VIII. Effects of adipic acid on growth and thermal resistance of spores of anaerobic sporeforming bacteria. Nippon Shokuhin Kogyo Gakkaishi. 1989; 36(7), 551–556. DOI: 10.3136/nskkk1962.36.7_551

[83] Nakae T, Nakanishi T. Application of steady state conduction system to microbiology Part V. Journal of the Agricultural Chemical Society of Japan. 1967; 41(9), 465–469. DOI:10.1271/nogeikagaku1924.41.9_465

[84] Gould WA. Understanding our past. Fundamentals of Food Processing and Technology. 1997; 9–20. DOI:10.1533/9781845696092.9

[85] Robertson GL. History of food packaging. Reference Module in Food Science. 2019; 1–49. DOI:10.1016/b978-0-08-100596-5.22535-3

[86] Hayakawa K, Ball CO. A note on theoretical heating curve of a cylindrical can of thermally conductive food. Canadian Institute of Food Technology

Journal. 1968; 1(2), 54–60. DOI:10.1016/s0008-3860(68)74464-0

[87] Bermudez A, Martinez A. A state constrained optimal control problem related to the sterilization of canned foods. Automatica. 1994; 30(2), 319–329. DOI:10.1016/0005-1098(94)90033-7

[88] Heldman DR. Introduction. Food Preservation Process Design. 2011; 1–18. DOI:10.1016/b978-0-12-372486-1.00001-4

[89] Mitchell EL. A review of aseptic processing. Advances in Food Research. 1988; 1–37. DOI:10.1016/s0065-2628(08)60284-0

[90] Mohamed IO. Computer simulation of food sterilization using an alternating direction implicit finite difference method. Journal of Food Engineering. 2003; 60(3), 301–306. DOI:10.1016/s0260-8774(03)00051-7

[91] Miranda-Zamora WR, Ludeña AL, Tapia DA, Bazán JF. Herramientas computacionales aplicadas a la evaluación de tratamientos térmicos de los alimentos envasados usando el método de Ball. Piura: UNP; 2010. 94.

[92] Miranda WR. Note: hyperbolic function or equation of the hyperbola? Research, March 2016, 11. DOI: 10.13140/RG.2.1.1239.5287

[93] Miranda-Zamora WR, Heldman DR. Diseño de procesos térmicos y alta presión de alimentos. Madrid: AMV; 2018. 634 p.

[94] Miranda-Zamora WR, Tucker GS. Procedimientos del tratamiento térmico de alimentos. Madrid: AMV; 2017. 290 p.

[95] Miranda-Zamora WR. Manual de tratamiento térmico y envasado de alimentos. Madrid: AMV; 2017. 622 p.

[96] Miranda-Zamora WR, Stoforos NG. Procesamiento térmico de alimentos teoría, práctica y cálculos. Madrid: AMV; 2016. 330 p.

[97] Miranda-Zamora WR, Vignolo TG, Leyva NL. Ingeniería del tratamiento térmico de alimentos. Piura: UNP; 2012. 268 p.

[98] Hayakawa K. Selective review of research results related to thermal process lethality estimation I. Lethality estimation and heat transfer. Japan Journal of Food Engineering. 2001; 2(1), 1–9. DOI:10.11301/jsfe2000.2.1

[99] Ball CO, Olson FCW. Sterilization in food technology theory, practice, and calculations. New York: McGraw-Hill; 1957. 654 p.

[100] Ball CO. Mathematical solution of problems on thermal processing of canned food. Berkeley: University of California Press; 1928. 245 p.

[101] Trystram G. Modelling of food and food processes. Journal of Food Engineering. 2012;110:269-277. DOI: 10.1016/j.jfoodeng.2011.05.001

[102] Augusto PED, Pinheiro TF, Cristianini M. Using Computational Fluid-Dynamics (CFD) for the evaluation of beer pasteurization: effect of orientation of cans. Ciência e Tecnologia de Alimentos. 2010;30(4): 980-986. DOI: 10.1590/S0101-20612010000400022

[103] Singh A, Singh AP, Ramaswamy HS. Computational techniques used in heat transfer studies on canned liquid-particulate mixtures. Trends in Food Science & Technology. 2015;43: 83-103. DOI: 10.1016/j.tifs.2015.02.001

[104] Lemus-Mondaca RA, Vega-Gálvez A, Moraga NO. Computational Simulation and Developments Applied to Food Thermal Processing. Food

Engineering Reviews. 2011;3:121-135. DOI: 10.1007/s12393-011-9040-x

[105] Martins RC. Simple finite volumes and finite elements procedures for food quality and safety simulations. Journal of Food Engineering. 2006;73:327-338. DOI: 10.1016/j.jfoodeng.2005.01.033

[106] Ho QT, Carmeliet J, Datta AK, Defraeye T, Delele MA, Herremans E, Opara L, Ramon H, Tijskens E, van der Sman R, Liedekerke PV, Verboven P, Nicolaï, BM. Multiscale modeling in food engineering. Journal of Food Engineering. 2013;114:289-291. DOI: 10.1016/j.jfoodeng.2012.08.019

[107] Erdogdu F, Sarghini F, Marra F. Mathematical Modeling for Virtualization in Food Processing. Food Engineering Reviews. 2017;9:295-313. DOI: 10.1007/s12393-017-9161-y

[108] Wang L, Sun DW. Recent developments in numerical modelling of heating and cooling processes in the food industry—a review. Trends in Food Science & Technology. 2003; 14: 408-423. DOI: 10.1016/S0924-2244(03) 00151-1

[109] Augusto PED, Cristianini M. Numerical Simulation of Packed Liquid Food Thermal Process Using Computational Fluid Dynamics (CFD). International Journal of Food Engineering. 2011;7(4):16. DOI: 10.2202/1556-3758.2418′

[110] Norton T, Sun DW. Computational fluid dynamics (CFD) -an effective and efficient design and analysis tool for the food industry: A review. Trends in Food Science & Technology. 2006;17: 600-620. DOI: 10.1016/j. tifs.2006.05.004

[111] Datta AK. Status of Physics-Based : Models in the Design of Food Products, Processes, and Equipment. Comprehensive Reviews in Food Science and Food Safety. 2008;7: 121-129. DOI: 10.1111/ j.1541-4337.2007.00030.x

[112] Verboven P, Defraeye T, Datta AK, Nocolai B. Digital twins of food process operations: the next step for food process models? Current Opinion in Food Science. 2020;35:79-87. DOI: 10.1016/j.cofs.2020.03.002

[113] Datta AK, Teixeira AA. Numerically predicted transient temperature and velocity profiles during natural convection heating of canned liquid foods. Journal of Food Science. 1988;53:191-195. DOI: 10.1111/ j.1365-2621.1988.tb10206.x

[114] Boz Z, Erdogdu Z. Evaluation of two-dimensional approach for computational modelling of heat and momentum transfer in liquid containing horizontal cans and experimental validation. Foods and Bioproducts Processing. 2013;91:37-45. DOI: 10.1016/ j.fbp.2012.08.005

[115] Lee MG, Yoon WB. Developing an effective method to determine the deviation of F value upon the location of a still can during convection heating using CFD and subzones. Journal of Food Process Engineering. 2014;37: 493-505. DOI: 10.1111/jfpe.12107

[116] Farazbakht F, Zamindar N, Hojjatoleslamy M, Toghraie D. Numerical simulation of transient heat transfer for tomato paste in semi rigid aluminum container. Journal of Food Measurement and Characterization. 2017;11:479-487. DOI 10.1007/ s11694-016-9415-z

[117] Malekjani N, Jafari SM. Simulation of food drying processes by Computational Fluid Dynamics (CFD); recent advances and approaches. Trends in Food Science & Technology. 2018;78: 206-223. DOI: 10.1016/j.tifs.2018.06.006

[118] Serami MS, Ramezan Y, Khashehchi M. CFD simulation and

experimental validation of in-container thermal processing in Fesenjan stew. Food Science & Nutrition. 2020;9:1079-1087. DOI: 10.1002/fsn3.2083

[119] Fadiji T, Coetzee CJ, Berry TM, Ambaw A, Opara UL. The efficacy of finite element analysis (FEA) as a design tool for food packaging: A review. Biosystems Engineering. 2018;174: 20-40. DOI: 10.1016/j.biosystemseng.2018.06.015

[120] Norton T, Tiwari B, Sun DW. Computational Fluid Dynamics in the Design and Analysis of Thermal Processes: A Review of Recent Advances. Critical Reviews in Food Science and Nutrition. 2013;53(3): 251-275. DOI: 10.1080/10408398.2010.518256

[121] Park HWP, Yoon WB. Computational Fluid Dynamics (CFD) Modelling and Application for Sterilization of Foods: A Review. Processes. 2018;6:62. DOI: 10.3390/pr6060062

[122] Boz Z, Erdogdu F, Tutar M. Effects of mesh refinement, time step size and numerical scheme on the computational modeling of temperature evolution during natural-convection heating. Journal of Food Engineering. 2014;123: 8-16. DOI: 10.1016/j.jfoodeng.2013.09.008

[123] Zhao CJ, Han JW, Yang XT, Qian JP, Fan BL. A review of computational fluid dynamics for forced-air cooling process. Applied Energy. 2016;168:314-331. DOI: 10.1016/j.apenergy.2016.01.101

[124] Erdogdu F, Karatas O, Sarghini F. A short update on heat transfer modelling for computational food processing in conventional and innovative processing. Current Opinion in Food Science. 2018;23:113-119. DOI: 10.1016/j.cofs.2018.10.003

[125] Shafiekhani S, Zamindar N, Hojatoleslami M, Toghraie D. Numerical simulation of transient temperature profiles for canned apple puree in semi-rigid aluminum based packaging during pasteurization. Journal of Food Science and Technology. 2016;53:2770-2778

[126] Rinaldi M, Malavasi M, Cordioli M, Barbanti D. Investigation of influence of container geometry and starch concentration on thermal treated in-package food models by means of Computational Fluid Dynamics (CFD). Food and Bioproducts Processing. 2018; 108:1-11. DOI: 10.1016/j.fbp.2017.12.003

[127] Kuriakose R, Anandharamakrishnan C. Computational fluid dynamics (CFD) applications in spray drying of food products. Trends in Food Science & Technology. 2010;21:383-398. DOI: 10.1016/j.tifs.2010.04.009

[128] Xia B, Sun DW. Applications of computational fluid dynamics (CFD) in the food industry: a review. Computers and Electronic in Agriculture. 2002;34 (1–3):5-24. DOI: 10.1016/S0168-1699(01)00177-6

[129] Nicolaï BM, Verboven P, Scheerlinck N. Modelling and simulation of thermal processes. In: Richardson P, editor. Thermal technologies in food processing. 1st ed. Cambridge: Woodhead Publishing; 2001. p. 91-112. DOI: 10.1533/9781855736610.2.91

[130] Lyczkowski WR. The Rise of the First Commercial CFD Codes: PHOENICS, FLUENT, FIDAP, CFX, FLOW-3D, and STAR-CD. In: Memoir P, editor. The History of Multiphase Science and Computational Fluid Dynamics. 1st ed. Cham: Springer; 2018. P. 185-187. DOI: 10.1007/978-3-319-66502-3_14

[131] Li B, Kang Z, Ma H. Research on the Meat Food Vacuum Cooling Model

based on ANSYS Simulation. Revista Ibérica de Sistemas e Tecnologias de Informação. 2016;E6: 184-196.

[132] Kızıltas S, Erdogdu F, Palazoglu TK. Simulation of heat transfer for solid–liquid food mixtures in cans and model validation under pasteurization conditions. Journal of Food Engineering. 2010;97:449-456. DOI:10.1016/j.jfoodeng.2009.10.042

[133] Ghani AGA, Farid MM, Chen XD, Richards P. Thermal sterilization of canned food in a 3-D pouch using computational ¯uid dynamics. Journal of Food Engineering. 2001;48(2):147-156. DOI: 10.1016/S0260-8774(00)00150-3

[134] Siriwattanayotin S, Yoovidhya T, Meepadung T, Ruenglertpanyakul W: Simulation of sterilization of canned liquid food using sucrose degradation as an indicator. Journal of Food Engineering. 2006;73(4):307-312. DOI: 10.1016/j.jfoodeng.2004.08.008

[135] Chourasia MK, Goswaki TK. CFD simulation of effects of operating parameters and product on heat transfer and moisture loss in the stack of bagged potatoes. Journal of Food Engineering. 2007;80:947-960. DOI: 10.1016/j.jfoodeng.2006.07.015

[136] Mirade PS. Prediction of the air velocity field in modern meat dryers using unsteady computational fluid dynamics (CFD) models. Journal of Food Engineering. 2003;60:41-48. DOI: 10.1016/S0260-8774(03)00009-8

[137] Stebel M, Smolka J, Palacz M, Adamczyk W, Piechnik E. Numerical investigation of the fluid flow distribution for the hydrofluidisation food freezing method. International Journal of Thermal Sciences. 2020;151: 106284. DOI: 10.1016/j.ijthermalsci.2020.106284

[138] Tattiyakul J, Rao MA, Datta AK. Simulation of heat transfer to a canned corn starch dispersion subjected to axial rotation. Chemical Engineering and Processing. 2001;40(4):391-399. DOI: 10.1016/S0255-2701(01)00116-7

[139] Tattiyakul J, Rao MA, Datta AK. Heat transfer to a canned corn starch dispersion under intermittent agitation. Journal of Food Engineering. 2002;54(4):321-329. DOI: 10.1016/S0260-8774(01)00218-7

[140] Okita WM, Reno MJ, Peres AP, Resende JV. Heat transfer computational fluid dynamics in the air blast freezing of guava pulp in large containers. Brazilian Journal of Chemical Engineering. 2013;30(4): 811-824. DOI: 10.1590/S0104-66322013000400013

[141] Malekjani N, Jafari SM. Food process modeling and optimization by response surface methodology (RSM). In: Sevda S, Singh A, editors. Mathematical and Statistical Applications in Food Engineering. 1st ed. Florida: CRC Press; 2020. p. 181-203. DOI: 10.1201/9780429436963-13

[142] Simpson R, Ramirez C, Jiménez D, Almonacid S, Nuñez H, Angulo A: Simultaneous multi-product sterilization: Revisited, explored, and optimized. 2019;241:149-158. DOI: 10.1016/j.jfoodeng.2018.08.007

[143] Alonso AA, Pitarch JL, Antelo LT, Vilas C. Event-based dynamic optimization for food thermal processing: High-quality food production under raw material variability. Food and Bioproducts Processing. 2021;127: 162-173. DOI: 10.1016/j.fbp.2021.02.013

[144] Abakarov A, Nuñez M. Thermal food processing optimization: Algorithms and software. Journal of Food Engineering. 2013;115(4):428-442. DOI: 10.1016/j.jfoodeng.2012.02.013

[145] Holdsworth SD, Simpson R. Thermal processing of packaged foods.

3rd edition. Cham: Springer; 2016. 516 p. DOI: 10.1007/978-3-319-24904-9

[146] Miri T, Tsoukala A, Bakalis S, Pistikopoulos EN, Rustem B, Fryer PJ. Global optimization of process conditions in batch thermal sterilization of food. Journal of Food Engineering. 2008;87(4):485-494. DOI: 10.1016/j.jfoodeng.2007.12.032

[147] Banga JR, Balsa-Canto E, Alonso AA. Quality and Safety Models and Optimization as Part of Computer-Integrated Manufacturing. Comprehensive Reviews in Food Science and Food Safety. 2008;7: 168-174. DOI: 10.1111/j.1541-4337.2007.00023.x

[148] Li J, Wang K, Gao Y, Ma C, Sun D, Hussain MA, Qayum A, Jiang Z, Hou J. Effect of thermal treatment and pressure on the characteristics of green soybean tofu and the optimization conditions of tofu processing by TOPSIS analysis. LWT – Food Science and Technology. 2021;136(1):110314. DOI: 10.1016/j.lwt.2020.110314

Chapter 8

Enhancement of Heat Transfer Using Taylor Vortices in Thermal Processing for Food Process Intensification

Hayato Masuda

Abstract

We are witnessing a transition from the traditional to novel processing technologies in the food industry to address the issues regarding energy, environment, food, and water resources. This chapter first introduces the concept of food process intensification based on vortex technologies to all food engineers/researchers. Thereafter, the novel processing methods for starch gelatinization/hydrolysis and heat sterilization based on Taylor–Couette flow are reviewed. In fluid mechanics communities, the Taylor–Couette flow is well-known as a flow between coaxial cylinders with the inner cylinder rotating. Recently, this unique flow has been applied in food processing. In starch processing, enhanced heat transfer through Taylor vortex flow significantly improves gelatinization. In addition, effective and moderate mixing leads to an increase in the reducing sugar yield. In sterilization processing, the enhanced heat transfer also intensifies the thermal destruction of *Clostridium botulinum*. However, a moderate heat transfer should be ensured because excessive heat transfer also induces thermal destruction of the nutritional components. The Taylor–Couette flow is only an example considered here. There are various flows that intensify the heat/mass transfer and mixing in food processing. It is expected that this chapter will stimulate the development of food processing based on fluid technologies, toward food process intensification.

Keywords: food process intensification, thermal processing, Taylor–Couette flow, starch hydrolysis, heat sterilization

1. Introduction

In manufacturing processes, including those specific to the food industry, sustainable development is necessary because there is a limit on the energy and other resources. To achieve this goal, chemical industries have considered process intensification (PI) that might result in a paradigm shift. Although the definition of PI is still under discussion, a dramatic reduction in the process size is one of the common goals. One of the methods to achieve size reduction is the enhancement of transport rates, such as momentum, heat, and mass. For example, Harvey [1] successfully showed that, in the ester saponification process, the apparatus size was reduced by one-tenth compared with a traditional batch reactor, using an oscillatory baffled reactor exhibiting an excellent mixing performance. Therefore, PI technologies would bring about innovation in all the manufacturing processes. In fact, the introduction of PI technologies has promoted various processes, for example, biopharmaceutical processes [2, 3]. The concept of PI should also be applied to food processing to establish energy/resource-saving processing. However, PI has not gained significant attention in the food industry. Boom et al. [4] analyzed three reasons for this: 1) food processing is largely based on traditional methods; 2) processing costs represent a small fraction of the total cost of food production, with the raw material representing the major portion of the total cost in most cases; and 3) the requirement of absolute food safety is a necessary obstacle to processing innovation. However, we should consider the transition from traditional food processing to novel processing by leveraging PI technologies, considering the environment, energy, and increasing population.

Few researchers have attempted to accomplish food process intensification by controlling fluid (liquid food) motion to enhance the mixing and heat/mass transfer. For example, Müller et al. [5] proposed a novel UV-C treatment device for juices based on the Dean vortex technology. Dean vortex flow occurs in a coiled tube owing to centrifugal instability [6]. They successfully showed that the Dean vortex flow promoted the inactivation of microorganisms because the fluid element is more frequently transported to the irradiation region through convective motion. Zhang et al. [7] successfully demonstrated the efficient manufacturing method of *Fuzhu* (also known as *Yuba*) through Rayleigh–Bénard convection. Rayleigh–Bénard convection is the flow between horizontal planes whose temperature at the lower plane is higher than that at the upper plane [8]. The driving cause underlying the Rayleigh–Bénard convection is the local distribution of the fluid density. Therefore, the novel concept based on fluid engineering has the potential for innovation in food processing.

In this chapter, aiming toward food process intensification, the application of a unique vortex flow between rotating cylinders (Taylor–Couette flow) to thermal processing is introduced.

2. Taylor–Couette flow

Taylor [9] first discovered and analyzed the unique vortex flow generated between cylinders with the inner cylinder rotating. This flow experiences several transitions with an increase in the rotational speed of the inner cylinder. The flow dynamics are characterized by the Reynolds number [$Re = \rho R_i \omega d / \eta$] in the circumferential direction, where ρ, R_i, ω, d, and η are the fluid density, inner cylinder radius, rotational speed of the inner cylinder, gap width, and fluid viscosity, respectively. At a relatively low Re, a Couette flow is observed with no pressure gradient in the flow direction. When Re exceeds a critical Re (Re_{cr}), toroidal vortices appear to be counter-rotating, spaced regularly along the axis, as shown in **Figure 1**. These vortices are called Taylor vortex flows. In addition, each vortex cell is called a Taylor cell. The value of Re_{cr} that depends on the radius ratio R_i/R_o, was theoretically derived by Taylor [9]. After the initial transition, as Re increases, the Taylor vortex flow cascadingly transitions to a singly periodic wavy vortex flow, quasi-periodic wavy vortex flow, and weakly turbulent wavy vortex flow [10, 11]. Finally, the flow develops into a fully turbulent vortex flow. The dynamics of the Taylor–Couette flow are interesting from the viewpoint of not only fluid mechanics, but also process engineering because this flow system has few advantageous characteristics as a reactor. First, mixing and heat/mass transfer are enhanced by the toroidal motion within the Taylor cells. Second, each Taylor cell is extruded through a single

DOI: http://dx.doi.org/10.5772/intechopen.99443

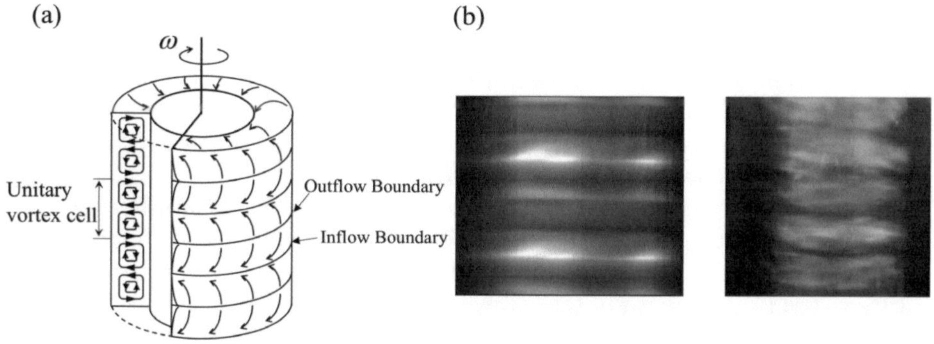

Figure 1.
Taylor–Couette flow: (a) schematic picture and (b) flow visualization. The left and right figures show the laminar Taylor vortex flow and wavy vortex flow, respectively. The black band corresponds to inflow boundaries.

file without breakdown when a small axial flow is added. In addition, mass transfer between the Taylor cells is prevented by an inward boundary where an inward secondary flow is formed. This implies that the axial dispersion that is a negative factor for uniform processing, is suppressed, while local mixing and heat/mass transfer are enhanced. Therefore, the continuous and uniform production is possible using this flow system as a reactor. Since Kataoka et al. [12] reported the excellent performance of Taylor–Couette flow in a chemical reactor, Taylor–Couette flow has been applied to various chemical processes, such as emulsion polymerization [13], photocatalytic reaction [14], particle synthesis [15], reverse osmosis [16], particle classification [17], and gas absorption [18]. According to these studies, the Taylor–Couette flow reactor has the potential to effectively intensify the processes compared with a traditional reactor, such as stirred tank reactor. In fact, among the chemical engineering communities, it is well-known that the Taylor–Couette reactor enables PI. Furthermore, few researchers have suggested that the Taylor–Couette flow apparatus is suitable for processes with shear-sensitive materials, such as food and bio-processes because the local strong shear force is absent. Haut et al. [19] applied the Taylor–Couette flow to the cultivation of CHO cells and reported the possibility of the Taylor–Couette flow-based bioreactor. To the best of our knowledge, the research conducted by Giordano et al. [20] is the first attempt of applying the Taylor–Couette flow in food processes. They showed that fructose–glucose isomerization could be efficiently conducted using a Taylor–Couette flow reactor. Subsequently, few researchers reported the excellent performance of Taylor–Couette flow in the non-thermal inactivation of bacteria in juice [21–23]. Although the application of Taylor–Couette flow to food processes is rather limited to non-thermal processing, efficient heat transfer by Taylor vortices should be utilized in thermal processing. Few research groups including the author have been trying to intensify thermal food processing based on Taylor–Couette flow. The intensification of starch processing and heat sterilization is introduced in this chapter.

3. Food process intensification using Taylor–Couette flow

3.1 Intensification of starch processing

Starch is typically a biopolymer that consists of 25% amylose (linear structure) and 75% amylopectin (branched structure). Detailed information on starch from the viewpoint of chemistry is reviewed in other articles [24]. Starch processing is frequently encountered in the manufacturing process of various types of food, such as beer, beverages, bread, and sauce. From a practical viewpoint, one of the most important types of starch processing is starch hydrolysis that comprises gelatinization, liquefaction, and saccharification. Enzymatic hydrolysis is described in this chapter because it is more prevalent than the other starch modifications, such as thermal and chemical treatment [25, 26]. In the starch hydrolysis process, the fluid viscosity intricately changes, as shown in **Figure 2**.

A significant increase in the viscosity was observed during gelatinization. Further, when enzyme (α-amylase) is added, the viscosity decreases as starch chains are broken down into glucose, maltose, maltotriose, and few higher oligomers. This intricate viscosity change is not favorable to the food engineers because the key operation is different between gelatinization and enzymatic liquefaction/saccharification processes. During gelatinization, heat transfer from the heated surface due to starch suspension and mass transfer between starch grains and water are required. In liquefaction/saccharification, highly efficient mixing of gelatinized starch and a small amount of enzyme is the most important operation. Therefore, individual apparatuses must be used. Consequently, the total size of starch hydrolysis process becomes large, as Baruque et al. indicated [28].

To make the total size compact, Baks et al. proposed the simultaneous and continuous processing of gelatinization and liquefaction/saccharification using an extruder [29, 30]. As shown in their studies, even at a high concentration of starch (600 g/L), gelatinization was completely conducted using the extruder. However, a high shear force was applied to the starch suspension in the extruder. This high shear force induces inactivation of the enzyme (α-amylase) [31, 32]. Therefore, other apparatuses such as stirred vessels are necessary for liquefaction/saccharification after gelatinization [30]. Paolucci-Jeanjean et al. [33] proposed a unique membrane reactor to conduct enzymatic hydrolysis using only one apparatus. However, the starch concentration was limited to 150 g/L because of the absence of mechanical agitation.

To intensify starch hydrolysis using a single apparatus, Masuda et al., Hubacz et al., and Matsumoto et al. applied a Taylor–Couette flow reactor for continuous starch hydrolysis [27, 34–39]. The features of the Taylor–Couette flow are described in the previous section. Taylor–Couette flow enhances not only mixing, but also heat/mass transfer. Therefore, it is expected that both gelatinization, where

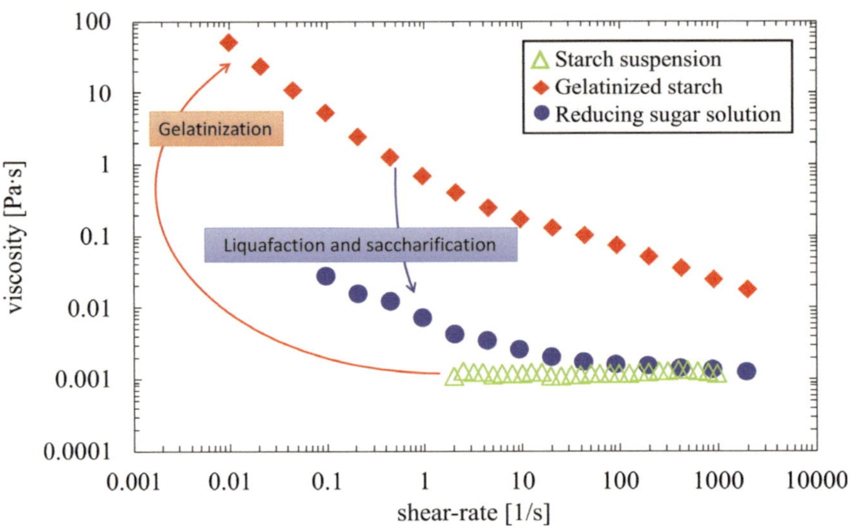

Figure 2.
Viscosity change at various shear rates during starch gelatinization/liquefaction/saccharification [27].

heat/mass transfer is necessary and liquefaction/saccharification, where mixing is necessary, are intensified using a single Taylor–Couette flow reactor.

As an example, a Taylor–Couette flow reactor utilized by Masuda et al. [35] is shown in **Figure 3**. The reactor consisted of a rotating inner cylinder, a stationary outer cylinder, and two divided water jackets. A starch suspension was introduced into the inlet. The enzyme was continuously fed using a syringe pump from the port set in the middle of the reactor. Therefore, the first and second half parts of the reactor were regarded as corresponding to the gelatinization and liquefaction/saccharification processes, respectively. High-temperature water was pumped in the first water jacket to promote gelatinization. Furthermore, moderate temperature water was pumped into the second water jacket to avoid the thermal deactivation of α-amylase.

The effects of Taylor vortices on starch gelatinization and hydrolysis were experimentally and numerically investigated in detail. **Figure 4** shows the impact of Taylor vortex flow on the degree of starch gelatinization (DSG). A high value of DSG was obtained when Taylor vortices were formed because the Taylor vortex flow enhanced the heat transfer from the heating surface. It should be noted that microscopic mass transfer around the starch granules was not considered in their simulation [36].

However, ascertaining whether Taylor vortices are formed within the reactor is not straightforward because the reactor is enwrapped in water jackets made of stainless steel. Therefore, to simulate the fluid flow in the reactor during starch gelatinization, Hubacz et al. [36] empirically established a mathematical model to describe the change in the rheological properties in response to gelatinization, as follows:

$$\eta = \frac{(0.0013DSG + 0.0031)}{1312^{n-1}}\dot{\gamma}^{n-1}, \tag{1}$$

where n [−] and $\dot{\gamma}$ [1/s] are the rheological model parameter and shear rate, respectively. **Figure 5** shows the axial velocity distribution near the inlet when the initial concentration of starch, C_0, is 50 g/L at the following values of ω: (a) 10 and (b) 22 rad/s. As clearly shown in **Figure 5**, at ω = 10 rad/s, it is confirmed that there

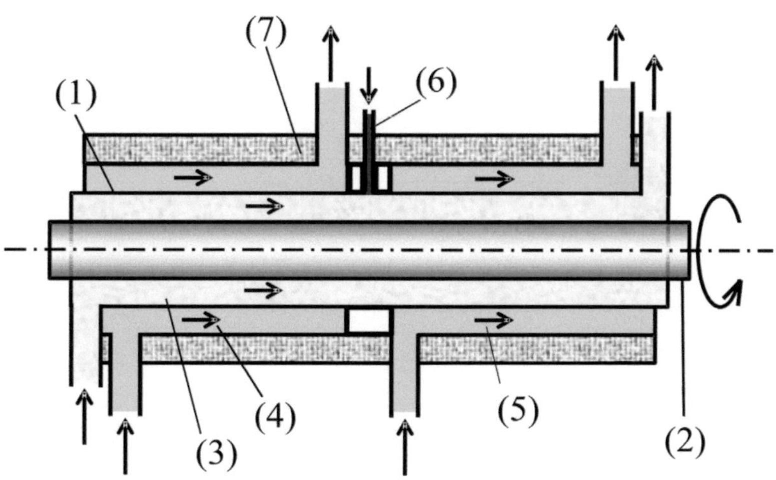

Figure 3.
Taylor–Couette flow reactor used for starch processing [35]: (1) stationary outer cylinder, (2) rotational inner cylinder, (3) starch suspension, (4) hot water for gelatinization, (5) variable temperature water for enzymatic reaction, (6) enzyme injection port, (7) insulator.

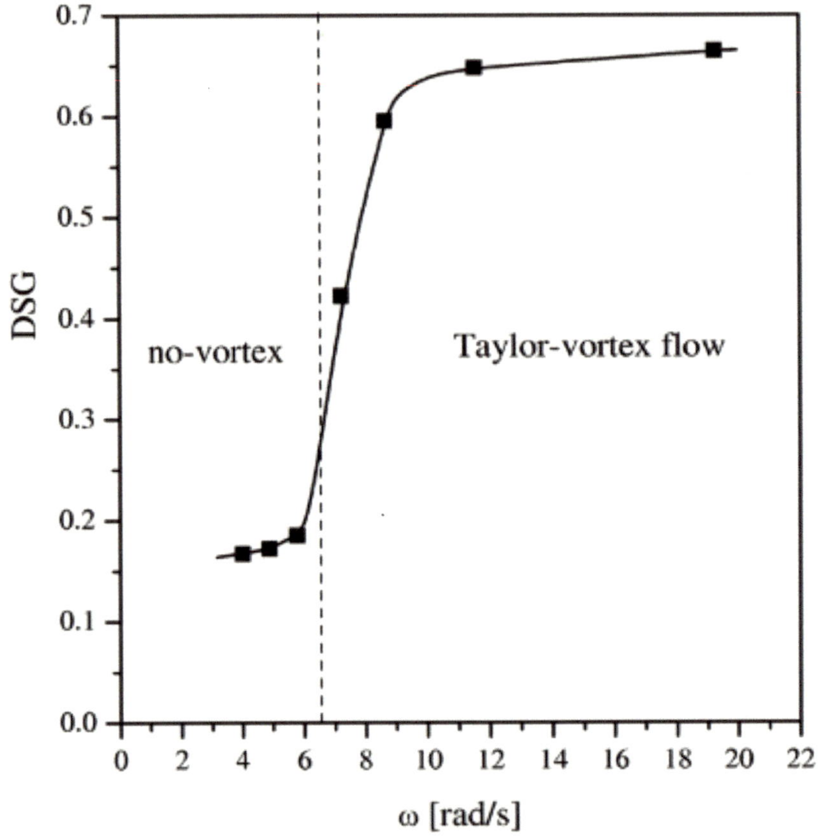

Figure 4.
Dependence of DSG (degree of starch gelatinization), obtained via two-dimensional simulation, on the rotational speed of inner cylinder (ω) [36]. The water jacket temperature, T_{hj}, *was assumed to be 65°C.*

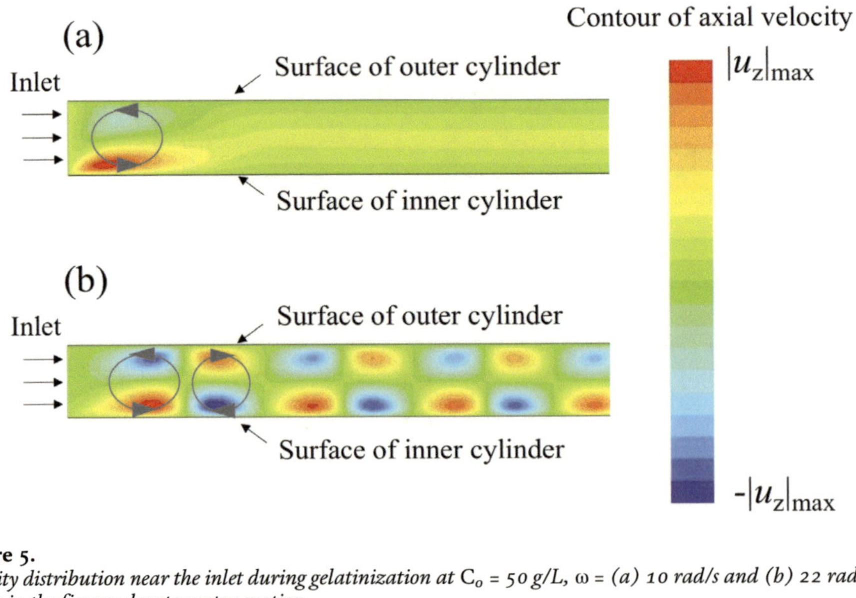

Figure 5.
Velocity distribution near the inlet during gelatinization at C_o *= 50 g/L, ω = (a) 10 rad/s and (b) 22 rad/s. The circles in the figures denote vortex motion.*

are no Taylor vortices, except near the inlet because of the lower centrifugal force. Therefore, the rheological model is reasonably advantageous for the practical design of the starch gelatinization process based on Taylor–Couette flow. In addition, as Van Den Einde et al. [40] indicated, starch granule degradation by thermomechanical treatment should also be considered. Hubacz et al. [36] confirmed that, as shown in **Figure 6**, there was no mechanical destruction of starch granules, and thermal degradation was not visible. Therefore, Taylor–Couette flow is suitable for the intensification of starch gelatinization owing to the efficient heat transfer without violent shear force.

The Taylor–Couette flow reactor intensifies starch gelatinization and liquefaction/saccharification. **Figure 7** shows that the relationship between the concentration of reducing sugar and effective Reynolds number at C_0 = 50, 150, and 300 g/L for the axial velocity u of 0.024 cm/s. It is noted that the flow condition was evaluated by the effective Reynolds number Re_{eff} because the apparent viscosity spatially changes due to the shear-thinning property of starch suspension. The detailed procedure for defining and calculating Re_{eff} is described in a paper by Masuda et al. [41]. As clearly shown in **Figure 7**, a higher yield of reducing sugar is obtained through the operation above Re_{cr} (dashed line in the figure) in all cases of C_0. Remarkably, starch is continuously and efficiently hydrolyzed even at relatively high concentrations of the starch suspension. The maximum yield is comparable to that obtained using a stirred batch reactor. Therefore, the conversion from batch to continuous is possible for food process intensification. However, a slight decrease in the yield was observed at higher Re_{eff} values. This decrease is explained by the axial

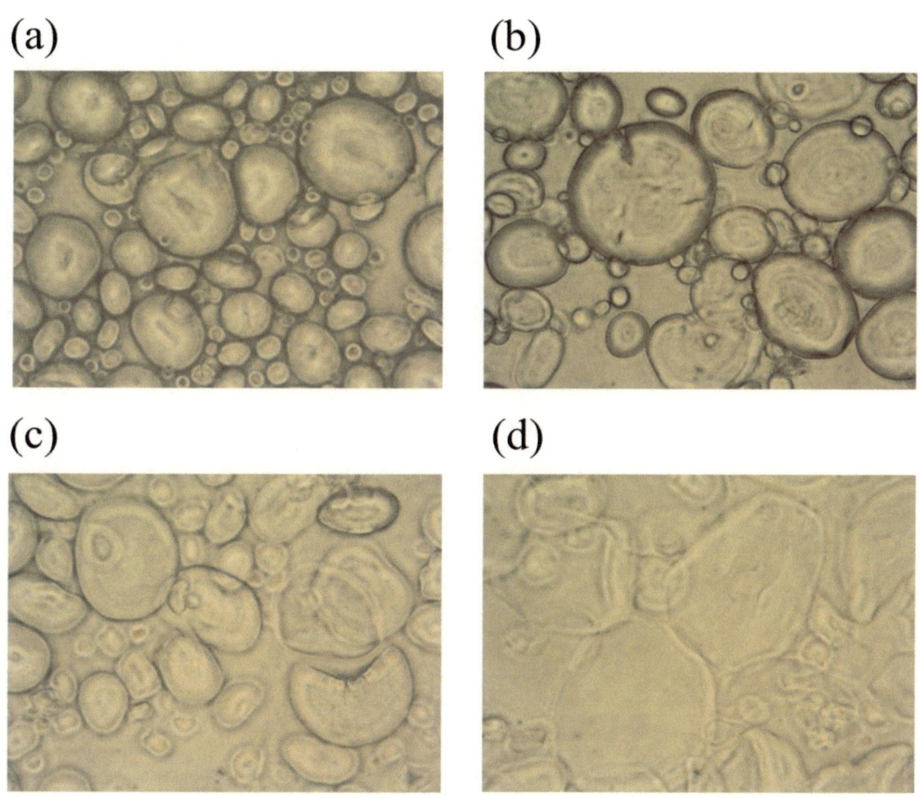

Figure 6.
Structure of starch observed using a light microscope: (a) native starch, gelatinized starch after treatment at (b) u = 0.099 cm/s, ω = 11.56 rad/s, T_{hj} = 60°C, (c) u = 0.099 cm/s, ω = 19.56 rad/s, T_{hj} = 65°C and (d) u = 0.099 cm/s, ω = 19.56 rad/s, T_{hj} = 85°C [36].

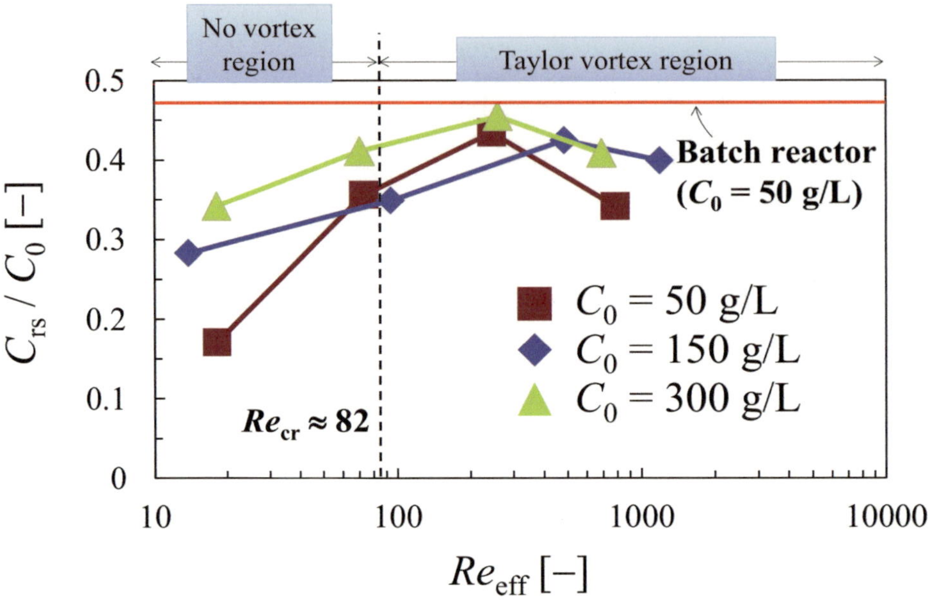

Figure 7.
Relation between the yield of reducing sugar (C_{rs}/C_0) and effective Reynolds number (Re_{eff}) at C_0 = 50, 150, 300 g/L, u = 0.024 cm/s, T_{hj} = 45°C [39]. The dashed line denotes the critical Re (Re_{cr}) where Taylor vortices are fully formed.

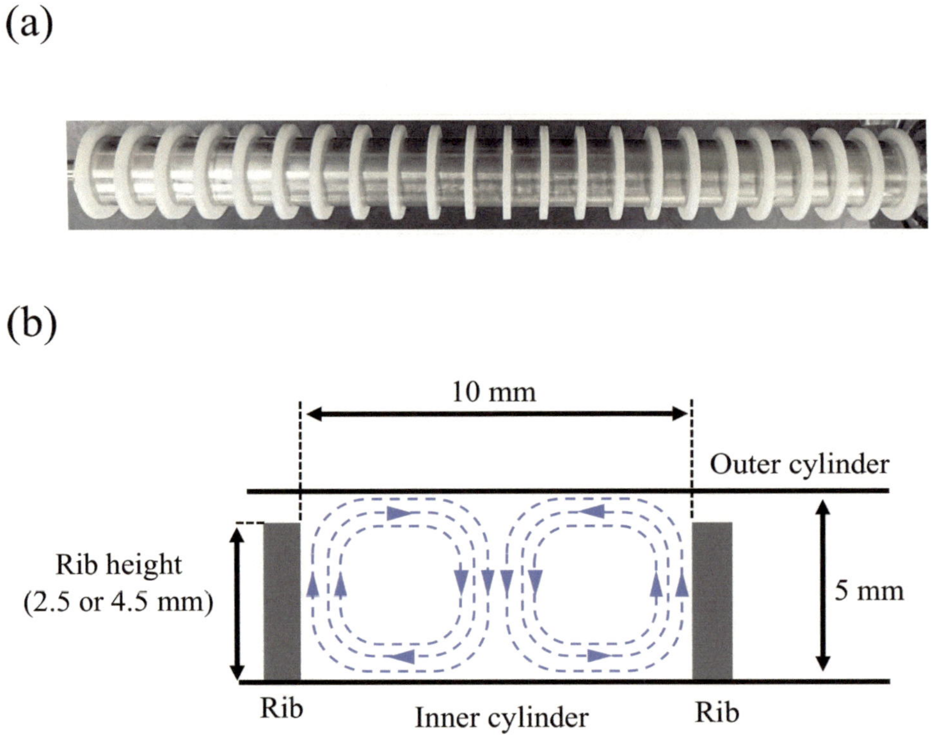

Figure 8.
Ribbed inner cylinder system [27]: (a) picture and (b) cross-sectional view of a pair of Taylor vortices between ribs.

dispersion and destabilization of the vortex structure during the enzymatic reaction [27]. At higher *Re*, a wavy motion is observed (called the wavy Taylor vortex flow). The wavy vortex flow significantly enhances mixing and heat/mass transfer within Taylor cells; furthermore, this also leads to axial dispersion through by-pass flow [42]. According to Richter et al. [43, 44], Taylor vortices can be stabilized and immobilized by a ribbed inner cylinder, as shown in **Figure 8**.

Consequently, the axial dispersion was suppressed even at a higher *Re*. Masuda et al. [27] successfully showed that the decrease in the yield at a higher Re_{eff} is suppressed by the equipment of ribs in the inner cylinder, as shown in **Figure 9**. Furthermore, as shown in **Figure 10**, the yield of small saccharides (glucose, maltose, and maltotriose) was significantly enhanced by utilizing the ribbed inner cylinder. This is because the ribbed inner cylinder enables the enhancement of mixing,

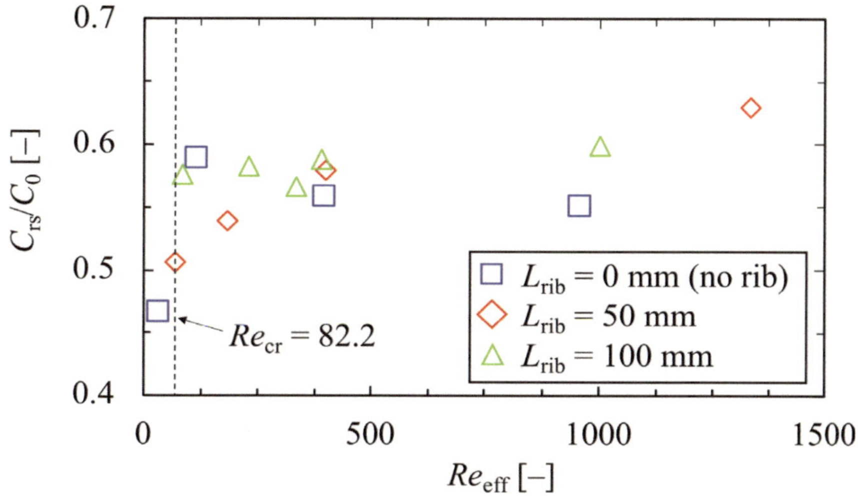

Figure 9.
Effect of Re_{eff} on C_{rs}/C_0 with three types of cylinders (L_{rib} = 0, 50, 100 mm) at u = 0.024 cm/s in starch hydrolysis experiments [27]. L_{rib} refers to the length of the ribbed section from the outlet.

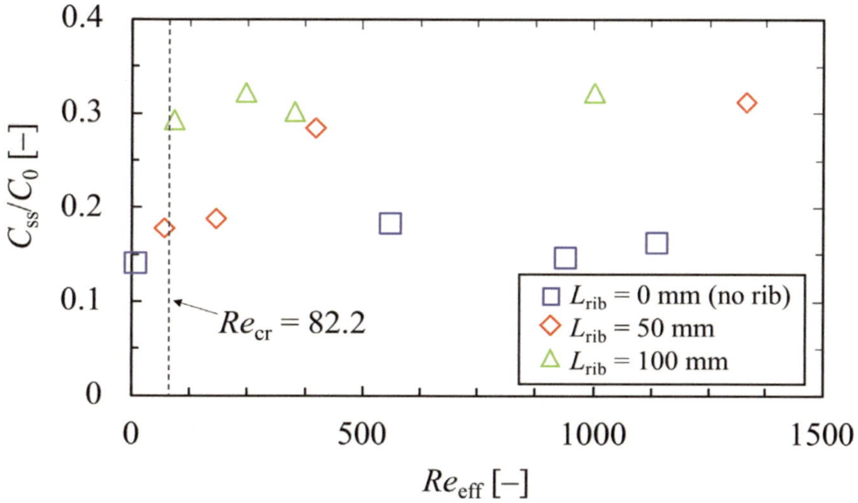

Figure 10.
Effect of C_{ss}/C_0 on Re_{eff} with three types of cylinders (L_{rib} = 0, 50, 100 mm) at u = 0.024 cm/s [27]. C_{ss} refers to small saccharide concentration.

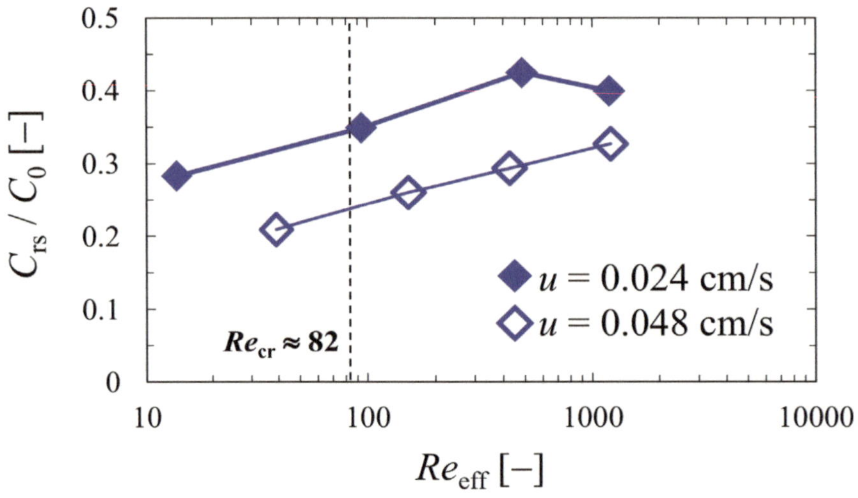

Figure 11.
Effect of axial velocity on C_{rs}/C_0 at C_0 = 150 g/L [39].

while the axial dispersion is suppressed at a higher *Re*. Finally, the effect of the axial velocity on the reducing sugar yield at C_0 = 150 g/L is shown in **Figure 11**.

At a higher axial velocity (u = 0.048 cm/s), the yield monotonically increases with Re_{eff} without a decrease at a higher Re_{eff}. Masuda et al. [39] explained that the transition from laminar Taylor vortex flow to wavy Taylor vortex flow occurs at a higher Re_{eff} than at a lower u because the axial flow enhances the stability of the Taylor vortex flow [45]. This should be investigated from the viewpoint of fluid mechanics. Nevertheless, the Taylor–Couette flow reactor promotes innovation in starch processing, for example, dramatic size reduction, high efficiency, and energy saving.

3.2 Intensification of heat sterilization processing

Heat sterilization is important for human health. Although novel technologies such as ultraviolet, ultrasonic, high-pressure, and cold plasma have been utilized [46], thermal sterilization plays a major role in the food industry. Recently, ohmic heating has recently been applied to heat sterilization processes [47]. However, the principle of scale-up for industries is under consideration. A traditional heat sterilizer, including a double-pipe, plate, and scrapped surface heat exchanger, faces problems such as clogging and high-pressure loss in the case of highly viscous liquid food. Therefore, heat sterilizers should be utilized for food process intensification. We consider the functions of an ideal sterilizer as follows:

1. High heat transfer performance in rapid heating;

2. Low shear force to avoid mechanical degradation of nutritional component;

3. Low pressure loss for saving energy.

These three functions are satisfied by adequately controlling the motion of liquid food. For example, chaotic advection and Dean vortex flow enable efficient and continuous heat sterilization [48, 49]. Taylor–Couette flow also offers a novel heat sterilization process. As described in the previous section, the Taylor–Couette flow offers efficient and moderate heat transfer. In addition, the shear-thinning properties of many liquid foods should be considered. Another advantage is that a lower power is required for pumping because the apparent viscosity decreases owing to the rotation of the inner cylinder. The Taylor–Couette flow sterilizer has the potential for food process intensification. Masuda et al. [50–52] numerically investigated the performance of a Taylor–Couette flow sterilizer. They assumed the sterilization process of highly viscous liquid food such as mayonnaise or ketchup, including the thermal destruction of the spores of *Clostridium botulinum* and the retention of thiamine.

Figure 12 shows the computational domain used in [51]. To eliminate the effect of back flows through Taylor vortex flow at the outlet, an extended section is imposed where the inner cylinder is stationary. This attempt does not affect the simulation results. They have solved the conservation equations of mass, momentum, heat, and chemical species, as follows [51]:

$$\nabla \cdot \mathbf{u} = 0, \tag{2}$$

$$(\mathbf{u} \cdot \nabla)\mathbf{u} = -\frac{\nabla p}{\rho} + \frac{1}{\rho}\nabla \cdot (2\eta \mathbf{D}) - \mathbf{g}\alpha(T - T_{\text{ref}}), \tag{3}$$

$$\mathbf{u} \cdot \nabla T = \frac{\lambda}{\rho C_{\text{p}}} \nabla^2 T, \tag{4}$$

$$\mathbf{u} \cdot \nabla C = \nabla \cdot (D_{\text{c}} \nabla C) + S, \tag{5}$$

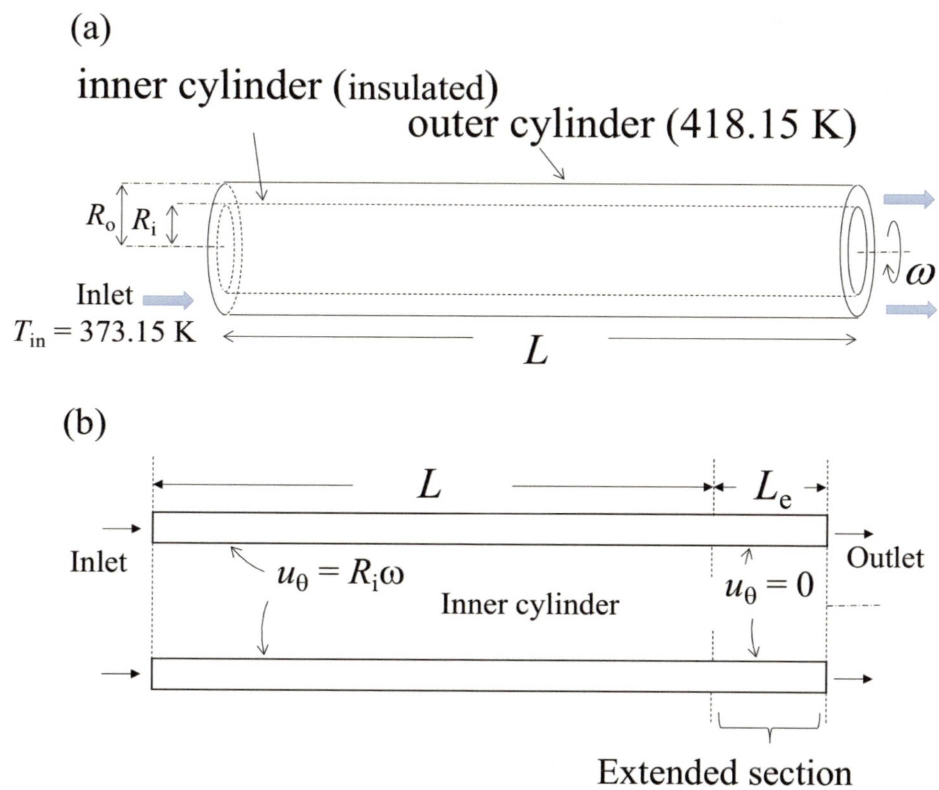

Figure 12.
Computational domain: (a) three-dimensional view without an extended section, (b) cross-sectional view with an extended section [51].

where **u** is the velocity, p is the pressure, ρ is the density, η is the viscosity depending on the shear rate, $\mathbf{D}$ (= $(\nabla \mathbf{u} + \nabla \mathbf{u}^T)/2$) is the rate of deformation tensor, **g** is the gravitational acceleration, α is the coefficient of volume expansion, T is the temperature, T_{ref} is the reference temperature, λ is the thermal conductivity, C_p is the specific heat capacity, C is the concentration, D_c is the diffusion coefficient, and S is the scalar source term. Because this simulation was assumed to be in a steady state, the time derivative term is omitted in Eqs. (2)–(5). It was assumed that the model fluid had a moderate shear-thinning property. According to Horak and Kessler [53], the thermal destruction of thiamine is followed by a second-order reaction model. The decrease in thiamine concentration due to destruction was included in the sink term, S, as shown in Eq. (5). Detailed information on the numerical procedure is described in [51]. The simulation code was validated.

Figure 13 shows the temperature distribution with the velocity vectors near the inlet at various values of Re_{eff}. In the case of Re_{eff} = 101.1 and 172.6 (**Figure 13(c)** and **(d)**), Taylor vortices were fully developed near the inlet, and consequently, heat transfer from the surface of the outer cylinder was significantly enhanced. This enhancement of the heat transfer is clearly confirmed from the bulk temperature distribution along the axis, as shown in **Figure 14**.

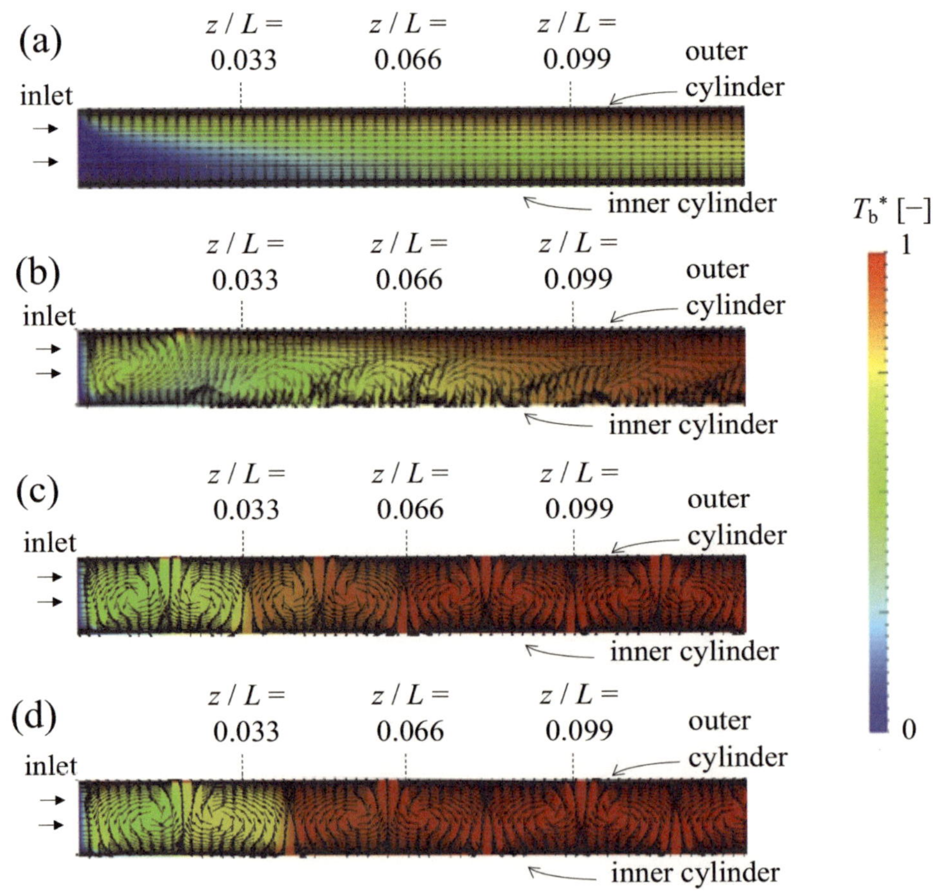

Figure 13.
Normalized bulk temperature distribution with velocity vectors in r-z plane near the inlet at (a) Re_{eff} = 0 (ω = 0 rad/s), (b) Re_{eff} = 43.7 (ω = 20 rad/s), (c) Re_{eff} = 101.1 (ω = 35 rad/s), (d) Re_{eff} = 172.6 (ω = 50 rad/s) [51].

To investigate the performance of heat sterilization, the equivalent lethality, F_0, was calculated from the temperature distribution. The value of F_0 is calculated as follows:

$$F_0 = \int_0^t \exp\left[\frac{E_a}{R}\left(\frac{1}{394.25} - \frac{1}{T(t)}\right)\right]dt, \tag{6}$$

where E_a is the activation energy for the destruction of *Clostridium botulinum*, and R is the gas constant. Finally, the local value of F_0 at an arbitrary axial position z is calculated as follows:

$$F_0(z) = \sum_{z=0}^{z} \Delta F_0 = \sum_{z=0}^{z} \Delta z \frac{dF_0}{dz}\bigg|_{\min}, \tag{7}$$

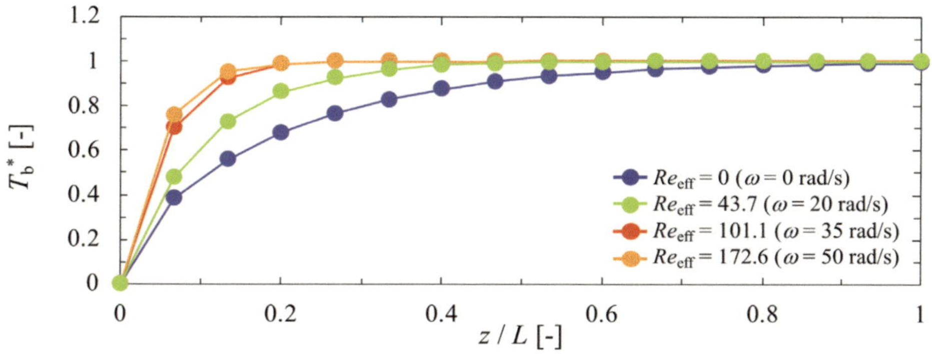

Figure 14.
Normalized bulk temperature distribution along the axis [51].

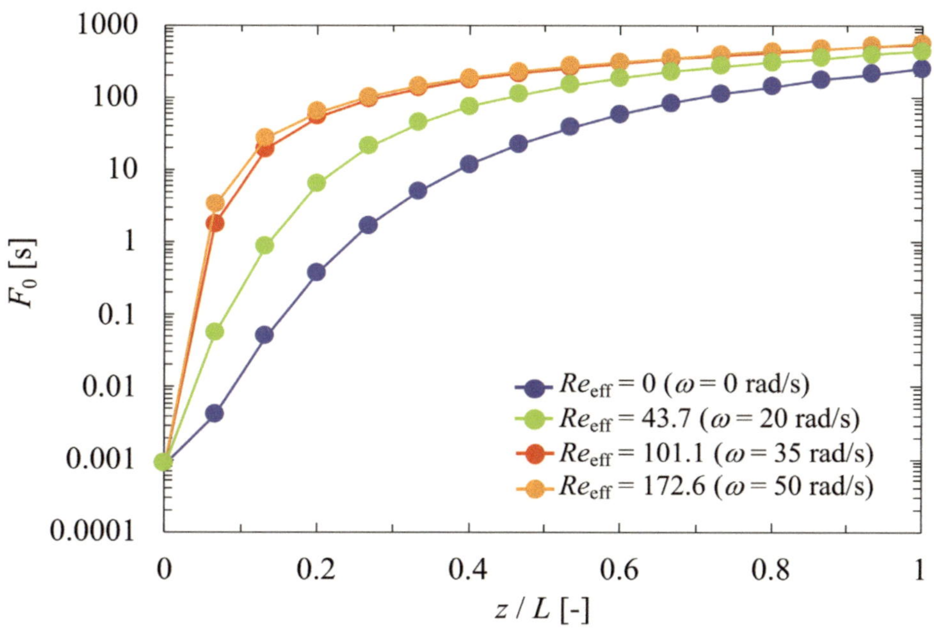

Figure 15.
Equivalent lethality distribution along the axis [51].

Figure 15 shows the axial distribution of F_0 along the axis. The significant increase in F_0 (higher than F_0 = 500 s) is observed under the condition at which Taylor vortices are developed (Re_{eff} = 101.1 and 172.6), as shown in **Figure 15**. This result indicates that Taylor–Couette flow has the potential to intensify the heat sterilization process. **Figure 16** shows the retention performance of thiamine during the sterilization process. Comparing the result at Re_{eff} = 101.1 with that at Re_{eff} = 172.6, it is confirmed from **Figure 15** that there is no clear difference in F_0. In addition, a clear difference in the thermal destruction of thiamine is not observed in **Figure 16**. Nevertheless, Ilo and Berghofer [54] indicated the mechanical destruction of thiamine by shear force. Therefore, the operation at Re_{eff} = 101.1, is preferable because of the lower shear force. It is valuable to investigate the effect of shear force on thiamine destruction in the future.

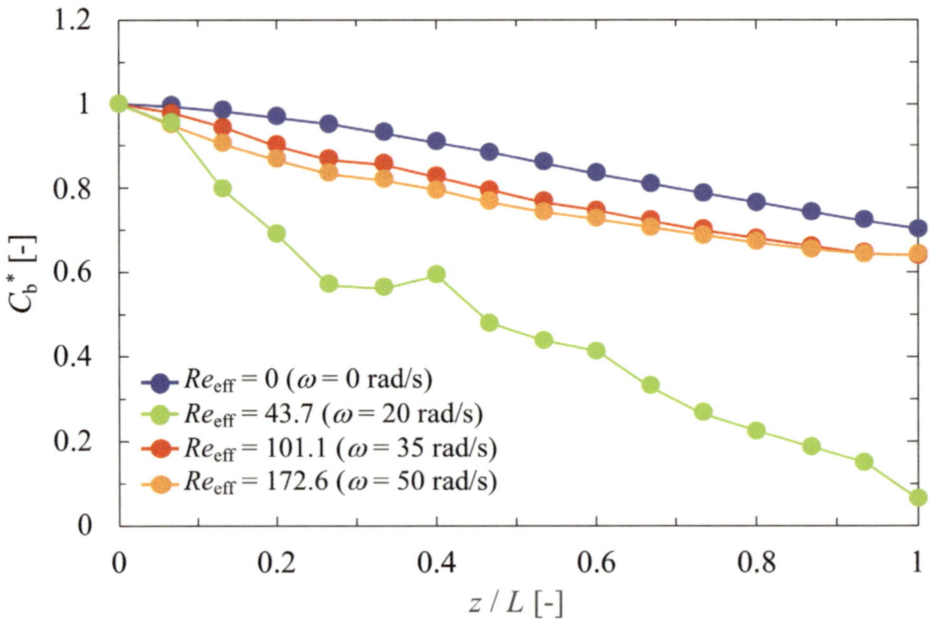

Figure 16.
Normalized thiamine concentration distribution along the axis [51].

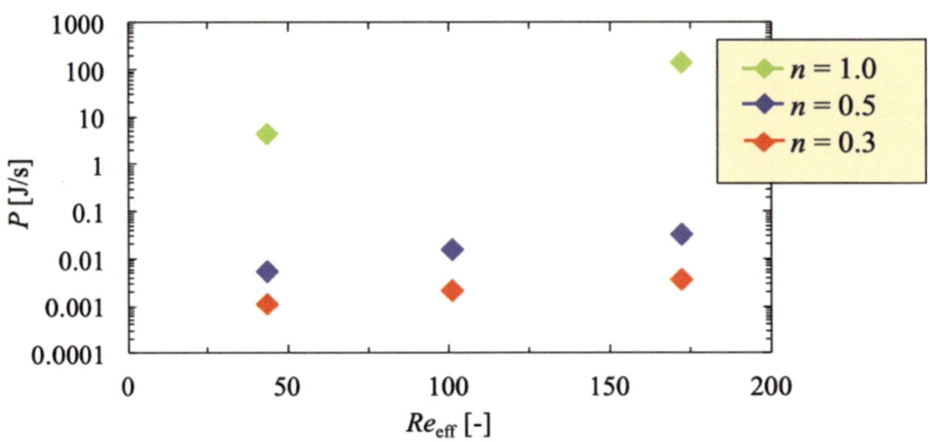

Figure 17.
Effect of power consumption on rheological properties in Taylor–Couette flow sterilizer.

Finally, the characteristics of energy consumption that are important for practical applications, are shown in **Figure 17**. In **Figure 17**, the energy consumption was calculated from the shear stress at the surface of the inner cylinder, as follows:

$$P = \omega R_{\mathrm{i}} \iint \tau_{\mathrm{r\theta}} \mathrm{d}A, \tag{8}$$

where $\tau_{r\theta}$ is the component of the shear stress tensor at the surface of the inner cylinder, and dA is the differential surface of the inner cylinder. It is noted that the value of n indicates the strength of the shear-thinning property. For Newtonian fluids, n corresponds to 1. Remarkably, **Figure 17** shows that the power consumption significantly decreases with an increase in the shear-thinning property because the apparent viscosity decreases owing to the shear force generated by the rotation of the inner cylinder. Therefore, the Taylor–Couette flow sterilizer enables energy-saving sterilization processing of liquid foods with shear-thinning properties.

4. Conclusions

In this chapter, novel food processing utilizing Taylor–Couette flow was introduced for food process intensification. As examples, starch processing and heat sterilization processes were specifically selected. With respect to starch processing, continuous and efficient gelatinization/liquefaction/saccharification were successfully conducted even in the case of high-concentration starch suspension. In addition, no clear thermal degradation of the starch granules was observed. Therefore, in the future, Taylor–Couette flow could be practically utilized in industries. In heat sterilization processing, enhancement of heat transfer by Taylor–Couette flow significantly improved the thermal destruction of *Clostridium botulinum*. Actually, the sufficient value of F_0 (higher than F_0 = 500 s) was obtained due to Taylor vortices. Based on the lethality, thermal destruction of nutritional components such as thiamine and mechanical destruction by shear force, the optimum operational conditions were proposed.

Taylor–Couette flow has the potential to intensify other processes as well. For example, an appropriate mixing performance of Taylor vortices would facilitate the manufacturing of sophisticated emulsions, such as multiple emulsions. Furthermore, other fluid techniques, such as chaotic advection, could incorporate novel processing. This chapter provides all food engineers with new insights into food process intensification.

Acknowledgements

Researches, introduced in this chapter, by the author was partially supported by JSPS KAKENHI (grant numbers JP18H03853, JP19KK0127, 20 K21110 and JP21K14450) and the Food Science Institute Foundation.

Conflict of interest

The author declares no conflict of interest.

Nomenclature

C	thiamine concentration [mg/L]
C_p	specific heat capacity [kJ/kg·K]
C_{rs}	reducing sugar concentration [g/L]
C_{ss}	small saccharide concentration [g/L]
C_0	initial concentration of starch [g/L]
$\mathbf{D}$	rate of deformation tensor [1/s]
d	gap width [−]
D_c	diffusion coefficient [m^2/s]
E	activation energy [kJ/mol]
E_a	activation energy for destruction of spores [kJ/mol]
F_0	lethality [s]
$\mathbf{g}$	gravity acceleration [m/s^2]
L	length of cylinders [m]
L_e	length of extended section of cylinders [m]
L_{rib}	length of ribbed section from outlet [mm]
n	power index [−]
R	gas constant [J/mol·K]
r	radial position [m]
Re	Reynolds number [−]
R_i	outer diameter of inner cylinder [m]
R_o	inner radius of outer cylinder [m]
S	scalar source term [mg/L·s]
T	temperature [K]
T_{hj}	heat jacket temperature [K]
$\mathbf{u}$	velocity [m/s]
u	axial velocity [m/s]
p	pressure [Pa]
z	axial position [m]

Greek letters

α	coefficient of volume expansion [1/K]
β	characteristic time [s]
$\dot{\gamma}$	shear-rate [1/s]
η	fluid viscosity [Pa·s]
η_0	zero shear rate viscosity [Pa·s]
λ	Thermal conductivity [W/m·K]
ρ	fluid density [kg/m^3]
τ	residence time [s]
ω	angular velocity of inner cylinder [rad/s]

Subscripts

b	bulk
cr	critical
eff	effective
ref	reference

Author details

Hayato Masuda
Department of Mechanical and Physical Engineering, Graduate School of Engineering, Osaka City University, Osaka, Japan,

*Address all correspondence to: hayato-masuda@eng.osaka-cu.ac.jp

References

[1] Harvey AP, Mackley MR, Stonestreet P. Operation and optimization of an oscillatory flow continuous reactor. Industrial & Engineering Chemistry Research. 2001; 40:5371-5377. DOI: 10.1021/ie0011223

[2] Shallan AI, Priest C. Microfluidic process intensification for synthesis and formulation in the pharmaceutical industry. Chemical Engineering and Processing - Process Intensification. 2019;142:107559. DOI: 10.1016/j.cep.2019.107559

[3] Satyawali Y, Vanbroekhoven K, Dejonghe W. Process intensification: The future for enzymatic processes?. Biochemical Engineering Journal. 2017; 121:196-223. DOI: 10.1016/j.bej.2017.01.016

[4] Boom RM, van der Goot AJ, Janssen AEM, Schroën KGPH. Food process intensification for better sustain-ability. In: Proceedings of 11th International Congress on Engineering and Food; 22-26 May 2011; Athnes; FMS1276

[5] Müller A, R. Stahl M, Graef V, M.A.P. Franz C, Huch M. UV-C treatment of juices to inactivate microorganisms using Dean vortex technology. Journal of Food Engineering. 2011;107:268-275. DOI: 10.1016/j.jfoodeng.2011.05.026

[6] Dean WR, Hurst JM. Note on the motion of fluid in a curved pipe. Mathematika. 1959;6:77-85. DOI: 10.1112/S0025579300001947

[7] Zhang J, Peng X, Guo S. Protein-lipid film (fuzhu) prepared from soymilk: Effects of soymilk convection on its formation, composition, and quality. LWT – Food Science and Technology. 2021;141:110909. DOI: 10.1016/j.lwt.2021.110909

[8] Bodenschatz E, Pesch W, Ahlers G. Recent developments in Rayleigh-Bénard convection. Annual Review of Fluid Mechanics. 2000;32:709-778. DOI: 10.1146/annurev.fluid.32.1.709

[9] Taylor GI. VIII. Stability of a viscous liquid contained between two rotating cylinders. Proceedings of the Royal Society A. 1923;223:289-343. DOI: 10.1098/rsta.1923.0008

[10] Di Prima RC, Swinney HL. Instabilities and transition in flow between concentric rotating cylinder. In: Swinney HL, Gollub JP, editors. Hydrodynamic Instabilities and the Transition to Turbulence. Berlin: Springer; 1981. p. 139-180

[11] Kataoka K. Taylor vortices and instabilities in circular Couette flows. In: Cheremisinoff, editor. Encyclopedia of Fluid Mechanics. Houston: Gulf Publications; 1986. p. 236-274

[12] Kataoka K, Doi H, Hongo T, Futagawa M. Ideal plug-flow properties of Taylor vortex flow. Journal of Chemical Engineering of Japan. 1975;8: 472-476. DOI:

[13] Kataoka K, Ohmura N, Kouzu M, Simamura Y, Okubo M. Emulsion polymerization of styrene in a continuous Taylor vortex flow reactor. Chemical Engineering Science. 1995;50: 1409-1416. DOI: 10.1016/0009-2509(94)00515-S

[14] Sczechowski JG, Koval CA, Noble RD. A Taylor vortex reactor for heterogeneous photocatalysis. Chemical Engineering Science. 1995;50: 3163-3173. DOI: 10.1016/0009-2509(95)00176-6

[15] Ogihara T, Matsuda G, Yanagawa T, Ogata N, Fujita K, Nomura M. Continuous synthesis of monodispersed silica particles using Couette-Taylor vortex flow. Journal of the Ceramic Society of Japan. 1995; 103:151-154. DOI:

10.2109/jcersj.103.151 10.2109/jcersj.103.151

[16] Lee S, Lueptow RM. Rotating reverse osmosis: a dynamic model for flux and rejection. Journal of Membrane Science. 2001;192:129-143. DOI: 10.1016/S0376-7388(01)00493-8

[17] Saomoto K, Horie T, Kumagai N, Takigawa T, Noui-Mehidi MN, Ohmura N. Dispersion of floating particles in a Taylor vortex flow reactor. Journal of Chemical Engineering of Japan. 2010;43:319-325. DOI: 10.1252/jcej.09We07

[18] Masuda H, Zheng W. Horie T, Ohmura N. Enhancement of gas hold-up with a Taylor vortex flow system equipped with ribs. Journal of Chemical Engineering of Japan. 2013;46:27-32. DOI: 10.1252/jcej.12we067

[19] Haut B, Ben Amor H, Coulon L, Jacquet A, Halloin V. Hydrodynamics and mass transfer in a Couette–Taylor bioreactor for the culture of animal cells. Chemical Engineering Science. 2003;58: 777-784. DOI: 10.1016/S0009-2509(02)00607-3

[20] Giordano RLC, Giordano RC, Cooney CL. Performance of a continuous Taylor–Couette–Poiseuille vortex flow enzymic reactor with suspended particles. Process Biochemistry. 2000;35:1093-1101. DOI: 10.1016/S0032-9592(00)00143-6

[21] Forney LJ, Pierson JA, Ye Z. Juice Irradiation with Taylor-Couette flow: UV inactivation of *Escherichia coli*. Journal of Food Protection. 2004;67: 2410-2415. DOI: 10.4315/0362-028X-67.11.2410

[22] Milly PJ, Toledo RT, Kerr WL, Armstead D. Hydrodynamic cavitation: characterization of a novel design with energy considerations for the inactivation of *Saccharomyces cerevisiae* in apple juice. Journal of Food Science. 2008;73:M298-M303. DOI: 10.1111/j.1750-3841.2008.00827.x

[23] Orlowska M, Koutchma T, Kostrzynska M, Tang J, Defelice C. Evaluation of mixing flow conditions to inactivate *Escherichia coli* in opaque liquids using pilot-scale Taylor–Couette UV unit. Journal of Food Engineering. 2014;120:100-109. DOI: 10.1016/j.jfoodeng.2013.07.020

[24] Ellis RP, Cochrane MP, Dale MFB, Duffus CM, Lynn A, Morrison IM, Prentice RDM, Swanston JS, Tiller SA. Starch production and industrial use. Journal of the Science of Food and Agriculture. 1998;77:289-311. DOI: 10.1002/(SICI)1097-0010(199807)77:3<289::AID-JSFA38>3.0.CO;2-D

[25] Tester RF, Debon SJJ. Annealing of starch — a review. International Journal of Biological Macromolecules. 2000;27: 1-12. DOI: 10.1016/S0141-8130(99)00121-X

[26] Tharanathan RN. Starch — Value addition by modification. Critical Reviews in Food Science and Nutrition. 2005;45:371-384. DOI: 10.1080/10408390590967702

[27] Matsumoto M, Masuda H, Hubacz R, Horie T, Iyota Y, Shimoyamada M, Ohmura N. Enzymatic starch hydrolysis performance of Taylor-Couette flow reactor with ribbed inner cylinder. Chemical Engineering Science. 2021;231: 116270. DOI: 10.1016/j.ces.2020.116270

[28] Baruque EA, Baruque MDA, Sant'Anna GL. Babassu coconut starch liquefaction: An industrial scale approach to improve conversion yield. Bioresource Technology. 2000;75:49-55. DOI: 10.1016/S0960-8524(00)00026-2

[29] Baks T, Ngene IS, van Soest JJG, Janssen AEM, Boom RM. Comparison of methods to determine the degree of gelatinization for both high and low

starch concentrations. Carbohydrate Polymers. 2007;67:481-490. DOI: 10.1016/j.carbpol.2006.06.016

[30] Baks T, Kappen FHJ, Janssen AEM, Boom RM. Towards an optimal process for gelatinisation and hydrolysis of highly concentrated starch–water mixtures with *alpha*-amylase from *B. licheniformis*. Journal of Cereal Science. 2008; 47:214-225. DOI: 10.1016/j.jcs.2007.03.011

[31] Özbek B, Yüceer S. α-Amylase inactivation during wheat starch hydrolysis process. Process Biochemistry. 2001;37:87-95. DOI: 10.1016/S0032-9592(01)00175-3

[32] van der Veen ME, Van Iersel DGV, van der Goot AJ. Shear-induced inactivation of α-amylase in a plain shear field. Biotechnology Progress. 2004;20:1140-1145. DOI: 10.1021/bp049976w

[33] Paolucci-Jeanjean D, Belleville MP, Rios GM, Zakhia N. Kinetics of continuous starch hydrolysis in a membrane reactor. Biochemical Engineering Journal. 2000;6:233-238. DOI: 10.1016/S1369-703X(00)00092-9

[34] Hubacz R, Buczyńska M. Starch gelatinisation in Couette-Taylor flow apparatus. Chemical and Process Engineering. 2011;4:267-279. DOI: 10.2478/v10176-011-0021-7

[35] Masuda H, Horie T, Hubacz R, Ohmura N. Process intensification of continuous starch hydrolysis with a Couette–Taylor flow reactor. Chemical Engineering Researh and Design. 2013; 91:2259-2264. DOI: 10.1016/j.cherd.2013.08.026

[36] Hubacz R, Ohmura N, Dłuska E. Intensification of Starch Processing Using Apparatus with Couette–Taylor Flow. Journal of Food Process Engineering. 2013;36:774-785. DOI: 10.1111/jfpe.12046

[37] Masuda H, Horie T, Hubacz R, Ohmura N. Numerical and experimental investigation of continuous starch hydrolysis with a Couette–Taylor flow reactor. In: Proceedings The 24th International Symposium on Transport Phenomena (ISTP 24) ; 1-5 November 2013; Yamaguchi. Japan; 2013. p. 402-407

[38] Hubacz R, Masuda H, Horie T, Ohmura N. Thermal treatment of starch slurry in Couette-Taylor flow apparatus. Chemical and Process Engineering. 2017;38:345-361. DOI: 10.1515/cpe-2017-0027

[39] Masuda H, Horie T, Hubacz R, Ohmura N, Shimoyamada M. Process development of starch hydrolysis using mixing characteristics of Taylor vortices. Bioscience, Biotechnology and Biochemistry. 2017;81:1-7. DOI: 10.1080/09168451.2017.1282806

[40] Van Den Einde RM, Van Der Goot AJ, Boom RM. Understanding molecular weight reduction of starch during heating-shearing processes. Journal of Food Science. 2008;68: 2396-2404. DOI: 10.1111/j.1365-2621.2003.tb07036.x

[41] Masuda H, Horie T, Hubacz R, Ohta M, Ohmura N. Prediction of onset of Taylor-Couette instability for shear-thinning fluids. Rheologica Acta. 2017; 56:73-84. DOI: 10.1007/s00397-016-0987-7

[42] Lueptow RM, Docter A, Min K. Stability of axial flow in an annulus with a rotating inner cylinder. Physics of Fluids. 1992;4:2446-2455. DOI: 10.1063/1.858485

[43] Richter O, Hoffmann H, Kraushaar-Czarnetzki B. Effect of the rotor shape on the mixing characteristics of a continuous flow Taylor-vortex reactor. Chemical Engineering Science. 2008;63: 3504-3513. DOI: 10.1016/j.ces.2008.04.003

[44] Richter O, Menges M, Kraushaar-Czarnetzki B. Investigation of mixing in a rotor shape modified Taylor-vortex reactor by the means of a chemical test reaction. Chemical Engineering Science. 2009;64:2384-2391. DOI: 10.1016/j.ces.2009.02.015

[45] Wereley ST, Lueptow RM. Velocity field for Taylor–Couette flow with an axial flow. Physics of Fluids. 1999;11: 3637-3649. DOI: 10.1063/1.870228

[46] Li X, Farid M. A review on recent development in non-conventional food sterilization technologies. Journal of Food Engineering. DOI: 10.1016/j.jfoodeng.2016.02.026

[47] Tiravibulsin C, Lorjaroenphon Y, Udompijitkul P, Kamonpatana P. Sterilization of coconut milk in flexible packages via ohmic-assisted thermal sterilizer. LWT. 2021;147:111552. DOI: 10.1016/j.lwt.2021.111552

[48] Kelder JDH, Ptasinski KJ, Kerkhof PJAM. Power-law foods in continuous coiled sterilisers. Chemical Engineering Science. 2002;57: 4605-4615. DOI: 10.1016/S0009-2509(02)00321-4

[49] Tian S, Barigou M. Using chaotic advection to enhance the continuous heat-hold-cool sterilisation process. Innovatice Food Science & Emerging Technologies. 2016;34:352-366. DOI: 10.1016/j.ifset.2016.03.007

[50] Masuda H, Horie T, Ohmura N, Shimoyamada M. Intensification of heat sterilization process for liquid foods using Taylor–Couette flow system. Chemical Engineering Transactions. 2017;57:1753-1758. DOI: 10.3303/CET1757293

[51] Masuda H, Hubacz R, Shimoyamada M, Ohmura N. Numerical simulation of sterilization process for shear-thinning food in Taylor–Couette flow system. Chemical Engineering & Technology. 2019; 42:859-866. DOI: 10.1002/ceat.201800600

[52] Masuda H, Hubacz R, Ohmura N, Shimoyamada M. Effect of rheological properties of liquid foods on heat sterilization process in Taylor–Couette flow apparatus. Chemical Engineering Transactions. 2019;75:31-36. DOI: 10.3303/CET1975006

[53] Horak F, Kessler H. Thermische Thiaminschädigung—Eine Reaktion 2. Ordnung. Zeitschrift Lebensmittal Untersuchung-und Forschung A. 1981; 173:1-6. DOI: 10.1007/BF01042831

[54] Ilo S, Berghofer E. Kinetics of thermomechanical destruction of thiamin during extrusion cooking. Journal of Food Science. 1998;63: 312-316. DOI: 10.1111/j.1365-2621.1998.tb15732.x

Chapter 9

Thoughts for Foods: Imaging Technology Opportunities for Monitoring and Measuring Food Quality

Ayman Eissa, Lajos Helyes, Elio Romano, Ahmed Albandary and Ayman Ibrahim

Abstract

In recent decades, the quality and safety of fruits, vegetables, cereals, meats, milk, and their derivatives from processed foods have become a serious issue for consumers in developed as well as developing countries. Undoubtedly, the traditional methods of inspecting and ensuring quality that depends on the human factor, some mechanical and chemical methods, have proven beyond any doubt their inability to achieve food quality and safety, and thus a failure to achieve food security. With growing attention on human health, the standards of food safety and quality are continuously being improved through advanced technology applications that depend on artificial intelligence tools to monitor the quality and safety of food. One of the most important of these applications is imaging technology. A brief discussion in this chapter on the utilize of multiple imaging systems based on all different bands of the electromagnetic spectrum as a principal source of various imaging systems. As well as methods of analyzing and reading images to build intelligence and non-destructive systems for monitoring and measuring the quality of foods.

Keywords: Food, Quality, Imaging technology, electromagnetic spectrum, image analysis

1. Introduction

The quality and safety of human food is the concern of all the elements of this system from the consumer, producers, and manufacturers of food, food control organizations, food safety organizations, and the market niche requirements. When agricultural and food products do not meet the quality standards and safety criteria, consumers lose faith in producers leading to the loss of these products' competitiveness in the market, and then significant economic loss. Although some systems are proposed to achieve food safety and quality by achieving a set of conditions that fall under the so-called Good Manufacturing Practices (GMP) and Hazard Analysis and Critical Control Point (HACCP) which represents the best way to achieve food security through all production steps. Unfortunately, with all these requirements for GMP and HACCP systems and others, they are not sufficient to ensure the production of safe food free of contaminants and defects, so it has become necessary to introduce modern technologies to quality inspect and detect blemishes and contamination. Therefore, the focus was on the development of non-destructive, modern, fast, reliable, and applicable methods that meet the needs of both food manufacturers and producers, as well as the desires of the consumer. In the present scientific climate, an acceleration in the growth of image processing technology has been observed among rapidly growing technologies. As well, image technology forms a core research area not only in engineering and computer science disciplines but also in the agricultural and food sectors. Image analysis is a study technique that aims to quantify the characteristics of each part of the image, both concerning their location and their filling. In fact, the densities are analytically observed as a function of the position concerning a reference point. The image is observed in its most basic units which are pixels (PICture ELement). They can have a square or rectangular shape and can take one or more values depending on the type of acquisition that is made, whether mono or multispectral. Even images taken with common cameras have more than one piece of information on each pixel. In fact, in this case, each pixel will have three numerical information with numbers ranging from 0 to 255, meaning by 0 the absence of color and with 255 the maximum intensity for that band of the electromagnetic spectrum. Image analysis is of interest in many areas of study, from the medical to the criminological, from the building to the agricultural sector. This diffusion is due to the fact that it is a very objective type of analysis that is based on a certain source (the image) and bases a series of calculations on it with a rigorous statistical approach. In the agri-food sector, image analysis is very successful because the appearance of a food product has a whole series of qualitative information that is difficult to parameterize by classical methods. Sensory approaches always remain invaluable tools of judgment, especially if conducted on a representative number of subjects and if carried out with an adequate plan of relief data and statistical adaptations. In addition, as far as the visual component is concerned, human vision, as already indicated, is limited to wavelengths between 390 and 700 nm, with greater sensitivity around 550 nm. The human eye is also influenced by the brightness of the background (simultaneous contrast effect) and tends to overestimate or underestimate information at the boundary between objects of different intensities (Mach Band effect) [1]. According to the widely accepted Retinex theory [2], there are three systems of independent cones consisting of receptors that read in three different wavelength regions of the visible spectrum.

Image analysis techniques, applied in the food field, show the following main advantages: objectivity, continuity over time, and rapid decision-making. An image is a representation of a two-dimensional or three-dimensional reality according to the independent spatial coordinates of an object. This is a one-plane transposition of the object's descriptive information, placed in exact positions concerning a reference point. In this context, [3] defined the image as a two-dimensional function, f (x, y), where x and y are the spatial coordinates and the amplitude of function, f at any pair of coordinates (x, y) is called the intensity or gray level of the image at that point as depicted in **Figure 1**. Consequently, if x,y, and the amplitude values of f are finite and discrete quantities, the image is defined as a digital image. So, the digital image is composed of a finite number of elements called pixels, each of which has a particular location and value. The images are generated by the combination of an energy source (electromagnetic, but also ultrasound) and the reflection of the energy emitted by the object.

A digital image is an ordered set of numbers each representing the intensity of reflection of the electromagnetic band of which the camera is sensitive. So, the

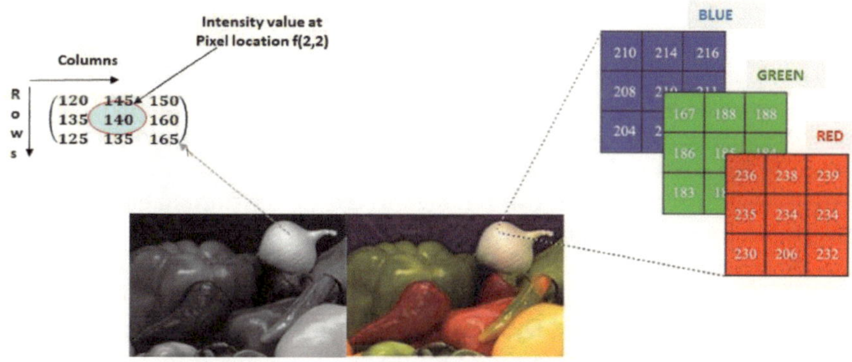

Figure 1.
Illustrates gray and color images: an array of pixels intensity and color values.

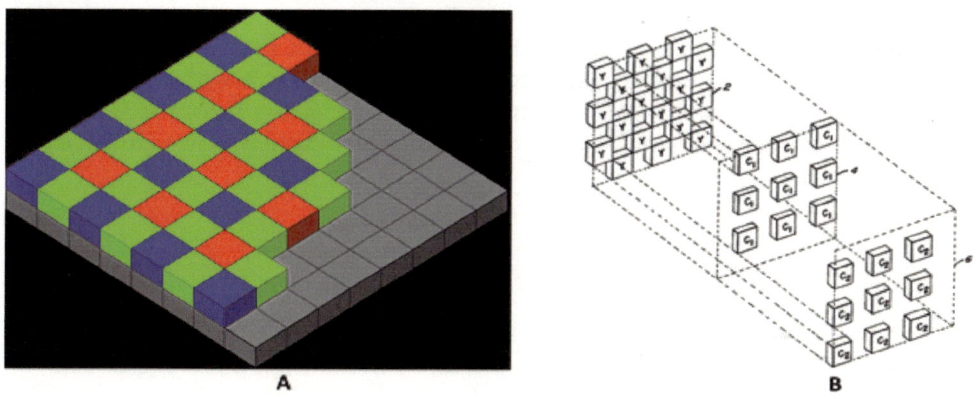

Figure 2.
Depicts an arrangement of arranged individual pixels, A) Bayer mosaic, and B) Debayering Process. (http://en.wikipedia.org/wiki/Image:US03971065_Bayer_Front.png).

digital image is composed of a finite number of elements called pixels, each of which has a particular location and value. Most digital cameras are equipped with sensors that allow the reading of individual pixels arranged in an arrangement called the Bayer mosaic [4] as shown in **Figure 2A**, consisting of blocks of four pixels, two green, one red, and one blue. Because each pixel is sensitive only to its own color, the end result is an image with scattered red, green, and blue dots. To achieve gradual tones and smooth photography, the processor or editing software must subsequently debayer **Figure 2B**. Because each pixel is filtered to record only one of the three colors, the data for each pixel cannot fully specify each of the red, green, and blue values on its own. To obtain a color image, various demosaicing algorithms can be used to interpolate a set of complete red, green, and blue values for each pixel. These algorithms use the surrounding pixels of the corresponding colors to estimate values for a particular pixel. Different algorithms that require various amounts of computing power result in final images of variable quality. This can be done inside the camera, producing a JPEG or TIFF image, or outside the camera using raw data directly from the sensor. Multispectral cameras are equipped with as many sensors as there are bands of the spectrum from which you want to get the reflected information.

Therefore, an image obtained from a multispectral camera will be a three-dimensional array consisting of as many matrices as there are observed bands. Each pixel in the array will have the value of the intensity of the amount reflected by the photographed object, for that band of the spectrum. The same principle of operation is the basis of spectrophotometers that generally consist of a light source, a lamp, which changes typology in case it is the analysis in the visible spectrum or UV and, in some specific instruments, infrared rays. While normal photography instruments typically limit themselves to capturing the intensity of reflectance of a scene or object for a limited number of spectral bands, corresponding to those needed to produce an image that can be interpreted by the human eye, hyperspectral cameras can capture, for each pixel, the entire spectral response in a wide, almost continuous range, depending on the type of camera itself. As a result, many measurements are available for extracting useful information. Hyperspectral images, therefore, collect a considerable amount of information from the same subject or surface, but at the same time require an important amount of interpretative commitment [5]. Common cameras, which represent the most widely used image capture tool and historically early means of capture, are also optimized to capture photons of light from the visible spectrum, and from the wavebands needed to build an image that can be interpreted by the human eye, providing very limited spectral information [6]. The peaks in the reflectance spectrum, detected by specific equipment, correspond to low absorption of the incident brightness and define the so-called "spectral signature", unique to each material [7]. Thus, the ability to exploit these differences in reflectance to characterize different materials through spectral response detection.

2. Image analysis procedures

The steps required in the analytical procedure in the case of the study of a phenomenon through image analysis do not differ much from those of classical analytical procedures. Indeed, both methodologies involve measurements, data processing, and consequent reporting. What fundamentally changes in the preparation of the support, which in the case of classical analysis consists in the preparation of the physical sample, while in the image analysis it involves the acquisition of the image from the physical sample, properly treated [8]. The elaboration phase involves the repetition of the measurements, their statistical evaluation, and the appropriate representation of the results. Image-analysis techniques allow you to simultaneously quantify multiple visual attributes and suggest criteria for classifying certain quality performance. The images obtained from the spectrograph, therefore, represent in number the same interval between the observed and available bands. Therefore, for each observed sample, as many photos will be available as there are spectrograph reading frequencies. Samples are hardly flat, and if you want to observe in their entirety, making sure that the reading is not invasive, they must be placed in the reading compartment, without making any cuts. Therefore, for example with potatoes or apples, there is a risk of having readings that can also depend on the variable distance from the emitted source of light. For this reason, images should be observed and sampled in areas where the distance from the sensor is equal for all samples [9]. So, the first step to take is to cut out the sub-sample of photography or better than more sub-photography samples in which to read the amount read by the sensor. Processing through R software can be developed through the EBImage package (Copyright © 2003–2021, Bioconductor), importing images through the read image function, then as arrays of values where pixels are identifiable through coordinates concerning a known point of origin [10]. This allows you to prepare shapes, usually regular (squares or rectangles), to crop subplots, through coordinate extraction. So, from each sub-plot, you can read the mean values and the standard deviation of the mean value. Therefore, in R, each sample

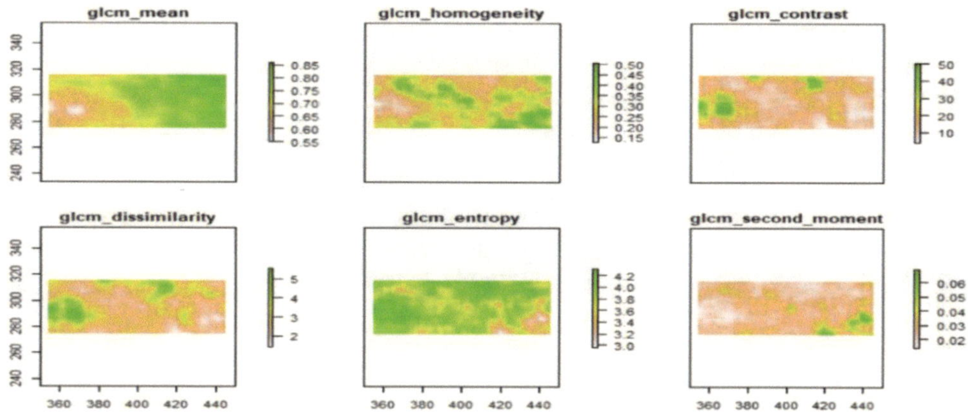

Figure 3.
GLCM on a sub-plot of a potato surface.

will constitute a vector that will be valued through the standard averages and deviations read from all the wavelengths used. So, the vectors of the samples will have as many values as there are wavelengths. If multiple sub-plots are used, they can be repeated for the same sample and can therefore be verified through ANOVA tests to assess the variance between the samples and that between the sub-samples. A similar image analysis path is possible even if there are few observed reflectance bands. In the case of RGB images, there are three bands (Red, Green, and Blue). In these cases, you can apply a type of analysis that relies on grayscale arrays (GLMC) [11–14]. The images are first rendered grayscale and the GLCM algorithm observes the relationship between each pixel and its neighbors **Figure 3**. In this way, parameters (homogeneity, contrast, dissimilarity, entropy) are measured that show whether the surface of the observed object is homogeneous or has irregularities.

3. Software available

In recent decades, many software, or additional packages specific to generic statistical software have been developed, such as Matlab, R, or Python [15]. Most algorithms refer to multi-way analysis, such as PARAFAC, PARAFAC2, N-PLS, Tucker3, and DTLD [16]. There are also some other open-source software implementations used for the multi-way analysis of other communities. For example, the tensor toolbox [17] is powerful for analyzing a wide type of tensor, these include dense, scattered, and symmetric tensors [18] and a Matlab tensorial decomposition package called tensor box that contains various algorithms optimized for the decomposition of a tensor, such as the fast dampened CP gauss-newton algorithm [19]. Recently, some multi-way analysis software packages running in the R environment have also been developed, such as ThreeWay [20] and multiway packages [21] developed by social science statisticians. Meanwhile, there are also some Python multi-way analysis packages available, such as Tensorly [22] and TensorD [23]. Other packages were born for other applications, such as EBImage or Image. Contour detector [24–26] refer to Otsu's algorithm [27], in an intention to discriminate bodies, backgrounds, particles and are currently used for image analysis in many sectors, including agro-industrial. An example of applying the identification of the contour and then the shape is shown in **Figure 4**, [13], which shows the reading of an image of a potato tuber.

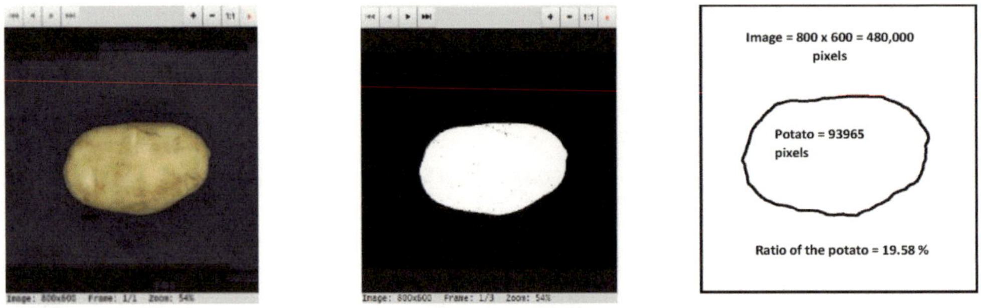

Figure 4.
Contour extraction on a potato RGB image.

4. Sources of digital images

Vision is the most important of the senses in human perception, so the images play the monocular most important role in enhancing this human perception. As a result of human limitations, where the human eye can identify and see objects in the visible light band in the electromagnetic spectrum. As well, in light of the evolution of consumers' desires for obtaining a high-quality and safe food product, and the inability of traditional methods to measure quality to meet the needs of the consumer, there were vigorous motives for developing imaging machines to cover almost the entire electromagnetic (EM) spectrum, ranging from gamma to radio waves. In order to be able to determine and measure the quality of food in general by identifying the appearance quality attributes, different phytochemical elements, internal structure, and detection of external and internal injuries and defects. The principal source for the images is the electromagnetic (EM) energy spectrum. Electromagnetic radiation is an electric and magnetic disturbance that propagates through space at the speed of light ($2.998 \times 108\ ms^{-1}$). The electromagnetic spectrum is the set of all possible frequencies or wavelengths of electromagnetic waves. Depending on their frequency or wavelength, electromagnetic waves interact differently with what they encounter in their propagation. The electromagnetic spectrum is the range of all frequencies of electromagnetic radiation from the shortest to the longest wavelength that can be generated physically. This range of wavelengths can be broadly divided into regions which include gamma rays, X-rays, ultraviolet, visible light, infrared, microwaves, and radio waves as shown in **Figure 5** [28, 29]. Electromagnetic radiation from the spectrum has found multiple applications

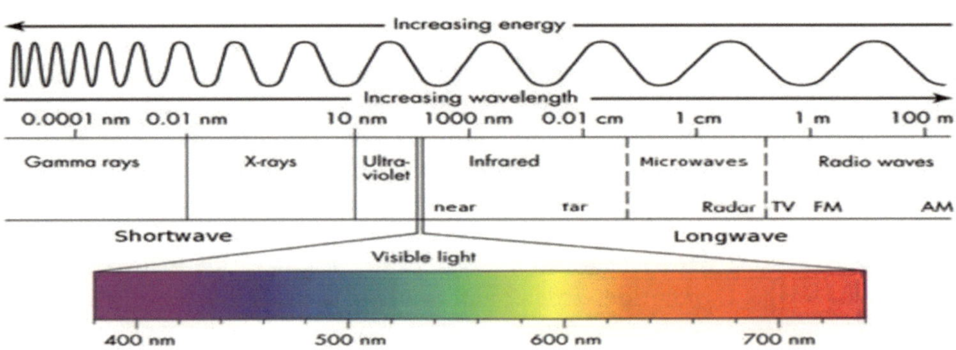

Figure 5.
Shows the different bands of electromagnetic (EM) spectrum.

ranging from communication to manufacturing. The following is a simplified explanation of all the different types of imaging according to each wavelength on the electromagnetic spectrum.

4.1 Gamma rays imaging

Gamma rays have the smallest wavelengths (wavelength: <0.01 nm) and the most energy of any wave in the electromagnetic spectrum, that they can pass through most types of materials as shown in **Figure 6**. This high penetration property makes gamma-ray imaging technology one of the most important imaging techniques for the internal properties of extremely thick objects. Gamma waves may be generated by nuclear explosions, lightning, accelerations of charged particles by strong magnetic fields, and the less dramatic activity of radioactive decay as mentioned [30]. Gamma decay occurs when a nucleus drops to a lower energy state from a higher energy state. Unlike alpha and beta decay, the chemical element does not change and carries no charge. The resulting emission produces gamma rays. The imaging of gamma-ray photons same as any band in the electromagnetic spectrum provides the ability to determine the origin of photons in space. Predominantly, the ability of gamma-ray imaging has been used in medical applications to trace specific radioactive markers to obtain information on transport, distribution, and metabolic or more specifically, to detect cancer or to study certain dynamical behavior, such as drug additions, and recently has been applied in astrophysics applications. Gamma-ray images capture by a Gamma camera (scintillation camera) which is an instrument developed for medical diagnostics to acquiring emitted gamma radiation from internal radioisotopes to create images and this process is called scintigraphy. Gamma camera consists of a detector, collimator, photomultiplier tubes PM tubes, preamplifier, amplifier, pulse height analyzer (PHA), X–Y positioning circuit, and display or recording device [31].

The detector, PM tubes, and amplifiers are housed in a unit called the detector head. The mechanism of growth and development of agricultural and food sectors requires a major development in modern technology to monitor agricultural operations in general, in addition to food production processes. Researchers have significantly improved the performance of a gamma ray-imaging camera, which is invisible to the human eye. The new technology has potential applications in scientific research, medical treatment, and environmental monitoring. In addition to the agricultural sector, many experiments have used multiple imaging techniques with gamma rays and have reached promising results for their practical application. In this regard, [32] mentioned many studies have been carried out in the last two

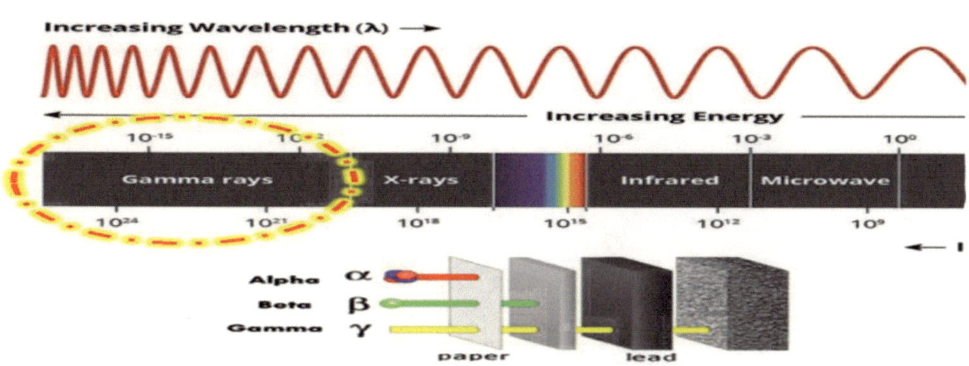

Figure 6.
Describe Gamma-ray position on the EM spectrum and the penetration property.

decades, using the gamma-ray computed tomography (CT) technique in several areas of knowledge other than medicine. As a result, used Gamma-ray computed tomography to characterize soil surface sealing and the study reached that the gamma-ray CT was able to confirm the occurrence of soil surface sealing due to the sewage sludge application and determine average densities and thickness of these layers. Through these results, concluded that the tool of gamma-ray CT allows a detailed analysis of soil bulk density profiles and the detection of very thin compacted or sealed layers. The gamma-ray computed tomography can be applied for wood density analysis, in the field for water infiltration studies and to provide information on the chemical composition of materials [33, 34]. Additionally, [30] pointed out the structure of the positron emission tomography (PET) scanners as depicted in **Figure** 7 where the PET scanners detect gamma rays with a ring of gamma-ray detectors placed around the subject. Where the special tracer molecules are ingested or injected into the living tissue. The main idea is focused on preparing the tracers by especially compounds to contain one or more radioactive atoms that spontaneously emit positrons (antimatters) positively charged electrons that rapidly colloid with electrons in the neighboring atoms. Then, the collision results in the annihilation of both the positron and electron and the creation of two gamma rays with the energy of a positron or electron. Furthermore, [35] mentioned that the positron-emitting tracer imaging system is one of the powerful techniques for researching the distribution and translocation of water, photoassimilate, mineral nutrients, and environmental pollutants to plants. This system works to detects two gamma rays produced by positron-emitting nuclides with a scintillation camera and therefore enables us to study the movement of elements in intact plants in real-time.

Accordingly, described the PET imaging system as a more compact system and flexible in the way to control the environment. Likewise, [36] explained that the positron-emitting tracer imaging system (PETIS) was developed to use the theory of PET in plants. It is equipped with a planar-type imaging apparatus and radioisotopes tracers such as ^{11}C, ^{13}N, ^{15}O, ^{52}Fe, ^{52}Mn, ^{64}Cu, and ^{107}Cd that are produced by a cyclotron and provides 2-D images. As well, [37] concluded that the 2-D and 3-D Gamma-ray imaging techniques have been successfully used in agriculture for the quantification and visualization of various compounds and mechanisms studies within plants such as water uptake and transportation, metal uptake, and transportation, photoassimilate translocation, etc. Modern technologies have become a necessary mechanism for growth and development in the field of agricultural production, food quality, and safety, and researchers are looking forward to

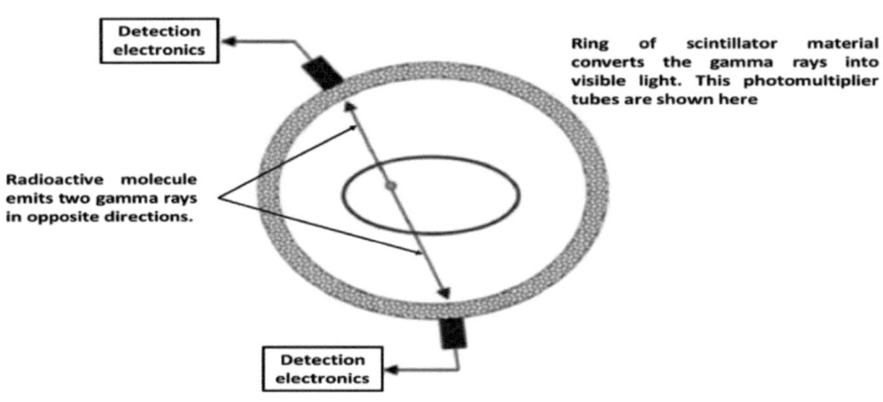

Figure 7.
Illustrative diagram of a positron emission tomography (PET) scan.

achieving the best results in achieving a sustainable development strategy. From this standpoint, gamma-ray imaging has been used with success in several fields, soil analysis, [38] mentioned that the current methods for soil sampling and lab analysis for soil sensing are time-consuming and expensive. So, used hyperspectral gamma-ray energy spectra to predict various surface and subsurface soil properties. It was concluded that the developed model provided a powerful prediction of clay, course, sand, and Fe contents in the 0–15 cm soil layer and pH and course sand contents in the 15–50 cm soil layer. Also, characterized and measured the mineral uptake and translocation within plants using positron emission tomography imaging system (PETIS) such as Mn in barley [39] describe the effects of the reduced form of glutathione (GSH) and study the behavior in the roots oilseeds rape plant [40]. Also, [41] applied PETIS to describe the absorption, transportation, and accumulation of cadmium from culture to spikelet in an intact rice plant. Furthermore, [42] studied using the ^{64}Cu as a tracer in the soybean plant for the transportation from root to the leaves and concluded that the ^{64}Cu could be a useful tracer for the use in plant studies such as the distribution and translocation of copper in intact plants using the PETIS as shown in **Figure 8**. Subsequently, [43] investigated the ability of the PETIS to visualize and quantitatively analysis of the real-time Cd dynamics from roots to grains in rice cultivars that differed in grain Cd concentrations using PETIS. Moreover, the utilize of positron emission tomography imaging system (PETIS) in the field of tracking water uptake and translocation within plants was studied. Where [35] studied the effect of Aminolevulinic acid (ALA) on $H_2^{15}O$ translocation from the roots to the shoots of rice plants in real-time by PETIS technology. As well, [44] applied PETIS technology to study the effect of light on $H_2^{15}O$ flow in rice plants. Where found that the plants were exposed to low light, the $H_2^{15}O$ flow was activated more slowly. By the same token, [45] studied the visualize of ^{15}O-water flow in tomato and rice plants in light and darkness by using PETIS technology. Although the applications of Gamma-ray imaging techniques are mainly used for research and development purposes, it has extremely great potential to serve as a tool for the development of several operations in the agriculture and food sectors.

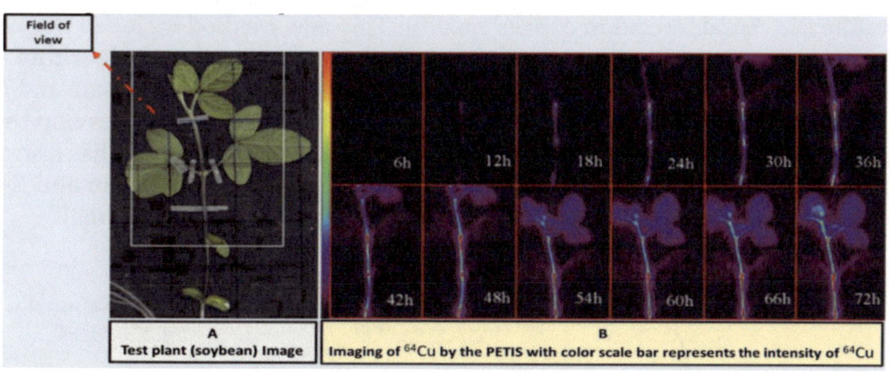

Figure 8.
Depicted the positron emission tomography imaging system setup for soybean.

4.2 X-rays imaging

X-rays are a kind of invisible electromagnetic energy with short wavelengths ranging from 0.01 to 10 nanometers and high frequency from 3×10^{19} to 3×10^{16} Hz, and thus high energies in the range 120 electron Volt (eV) to 120 kilo-electron Volt (keV), and it falls in the range of the electromagnetic (EM) spectrum between ultraviolet radiation and gamma rays. X-rays are short electromagnetic waves that behave like particles while interacting with the matter as discrete bundles of energy and are called photons or quanta. Almost, X-rays are classified into soft X-rays and hard X-rays. Soft X-rays have relatively short wavelengths of about 10 nanometers, while hard X-rays have wavelengths of about 100 picometers [46] as shown in **Figure 9**. In general, [3] mentioned that X-rays are among the oldest sources of EM radiation used for imaging. The best-known use of X-rays in medical diagnostics, but they also are used extensively in industry and other areas, like astronomy. Besides medical diagnostics imaging and astronomy, there are other applications of X-rays such as checking luggage at the airport, inspecting industrial ingredients, and security.

As a result of the powerful penetrating X-ray, it has become one of the most important modern applications used in the inspection of agricultural products and food in general. X-ray imaging techniques are the least used in non-destructive methods for internal quality evaluation which are gaining popularity nowadays in various fields of agriculture and food quality evaluation. Although, X-ray techniques, so far predominantly used in medical applications, but also have been explored for internal quality inspection of several agricultural products non-destructively when quality attributes are invisible on the surface of the products. Given considerations of product safety, consumer health, and meeting market needs, the non-destructive nature of these techniques has great potential for wide applications on agricultural and food products. In short, the action idea of an X-ray imaging system is based on the principle of transmission imaging technique, that the X-ray beam penetrates the object and attenuates based on the density variance of the object. Then this attenuated energy that passed through the object is identified through a photodetector, a film, or an ionization chamber on the other side. Thus, the attenuation coefficients of the object components lead to different contrast between these components [46–50]. Accordingly, [51] reported that the soft X-ray method was rapid and took only 3–5 s to produce an X-ray image. Undoubtedly, X-ray inspection systems are becoming one of the best solutions to ensure product quality, safety and prevent risks in the food sector. Based on the combination of multispectral and X-ray imaging technologies [52] presented a new method for automatic characterization of seed quality. This new method included the application of a normalized canonical discriminant analyses (nCDA) algorithm to obtain spatial and spectral patterns on different seed lots. Reflectance data and X-ray classes based on linear discriminant analysis (LDA) were used to develop the classification models. Concluded that multispectral and X-ray imaging has a strong relationship with seed physiological performance. Reflectance at 940 nm and X-ray data showed high accuracy (>0.96) to predict quality traits such as normal

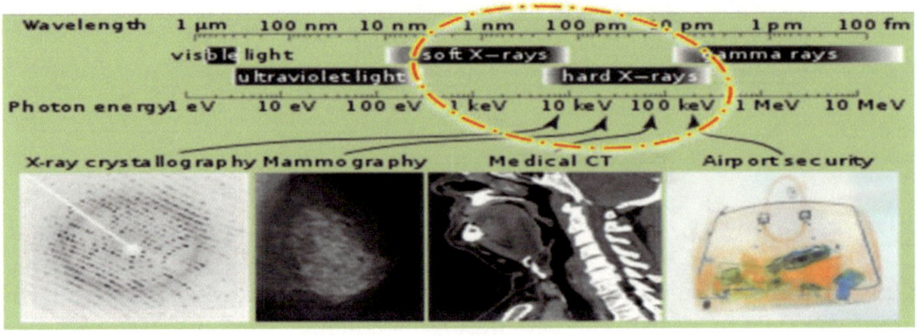

Figure 9.
Determines the location of X-radiation on the electromagnetic spectrum.

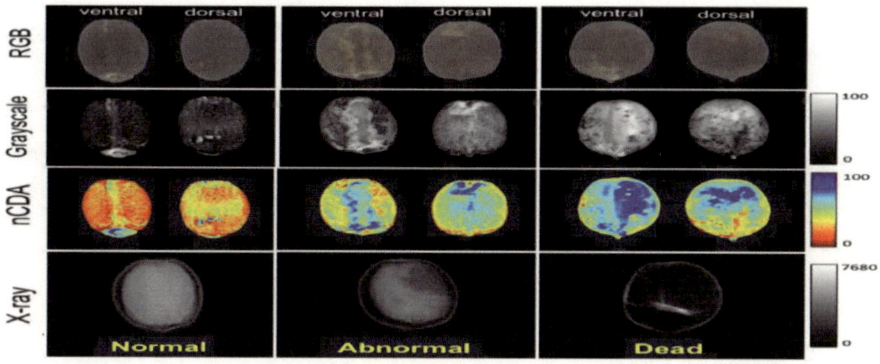

Figure 10.
Illustrates raw RGB images, reflectance images captured at 940 nm (grayscale and transformed images using nCDA algorithm), and X-ray images of ventral and dorsal surfaces of Jatropha curcas seeds.

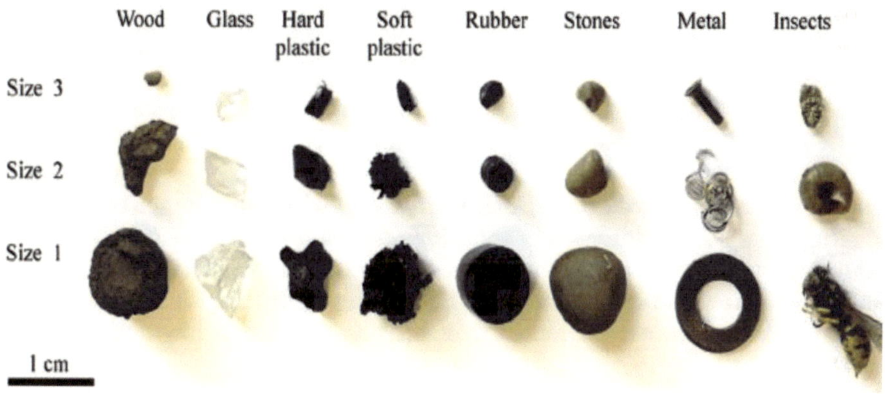

Figure 11.
Describes eight different foreign objects in three size groups.

seedlings, abnormal seedlings, and dead seeds as shown in **Figure 10**. These techniques can be alternative methods for rapid, efficient, sustainable, and non-destructive characterization of seed quality in the future, overcoming the intrinsic subjectivity of the conventional seed quality analysis. In a serious study for nondestructive inspection and detection of foreign materials in food products, [53] demonstrate a method for novelty detection of foreign objects **Figure 11** such as wood chips, insects, and soft plastics in food products using grating-based multimodal X-ray imaging. Through using X-ray imaging technique with three modalities absorption, phase contrast, and dark field to pixel correspondence and enhancing organic materials such as wood chips, insects, and soft plastics not detectable by conventional X-ray absorption radiography.

An example of X-ray images obtained of all food products with foreign objects from size group 2 at absorption, contrast, and dark field, from top to bottom, respectively as shown in **Figure 12**. It is clearly visible that there is a different contrast between the three imaging modalities. Concluded that the results give a clear indication of superior detection results from the grating-based method, and especially show promising detection results of organic materials. At the same time, [54] used the X-ray Imaging technique in a study conducted to detect the Infestation by Saw-Toothed Beetles of stored dates fruits, where its main goal was to investigate the capability of X-ray imaging in detecting internal infestations caused by the saw-toothed beetle in stored date fruits.

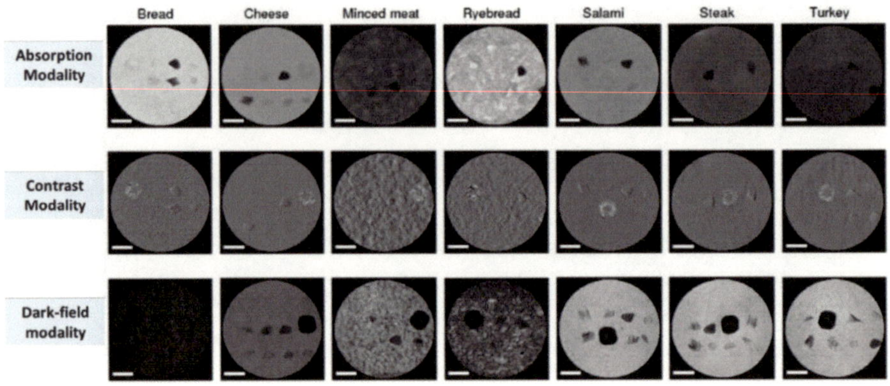

Figure 12.
Shows the X-ray images obtained for seven food products with foreign materials from size group 2, and the white bar represents 1 cm.

X-ray images of the dates were acquired at 40 kV potential and 1.6 mAs with a resolution of 512 × 512 by using an X-ray machine as shown in **Figure 13**. In the final analysis, the X-ray imaging system yielded around 97% accuracy in detecting internal infestation of dates with an adult beetle while using a pairwise classification method. Similarly, [55] presented an approach for visual detection of organic foreign objects such as paper and insects in food products using X-ray dark-field imaging. The results proved that the dark-field modality gave larger contrast-to-noise ratios than absorption radiography for organic foreign objects. Additionally, [56] developed an adaptive X-ray image segmentation algorithm based on the local pixels intensities and an unsupervised thresholding algorithm for the determination of infestation sites of several types of fruit such as citrus, peach, guava, etc.

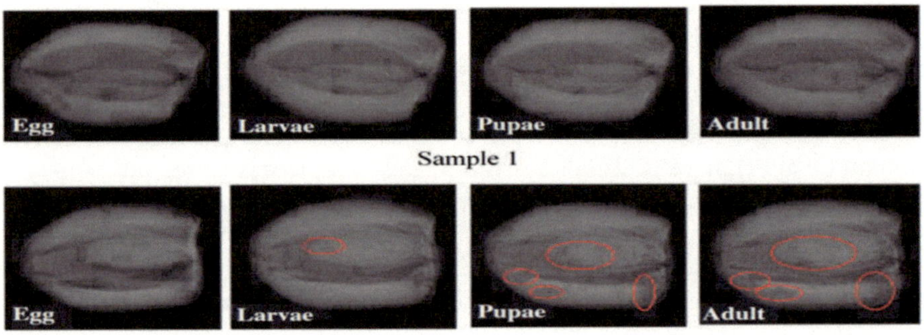

Figure 13.
Shows X-ray images of two dates infested.

The X-ray images were acquired through an X-ray imaging system which consists of a microfocus X-ray source, a line-scan sensor camera both of which are controlled by a desktop computer, and a frame grabber board to acquire and transfer the signal from the line-scan sensor to the host computer as shown in **Figure 14**. The developed algorithm proved fast in computation time and was implemented in the X-ray scanner for real-time quarantine inspection at a scanning rate of 1.2 m/min.

Thus, suspected sites of infestation inside the fruit can be accurately marked on the acquired X-ray image to aid the quarantine officer during the inspection **Figure 15** for guava and peach. In conclusion, the detection accuracies of the infestation detection experiments for guava and peach fruits were apparently

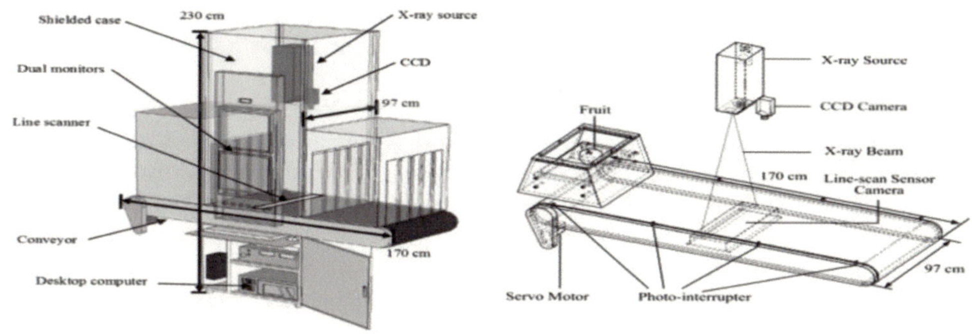

Figure 14.
Schematic drawing of the X-ray imaging system.

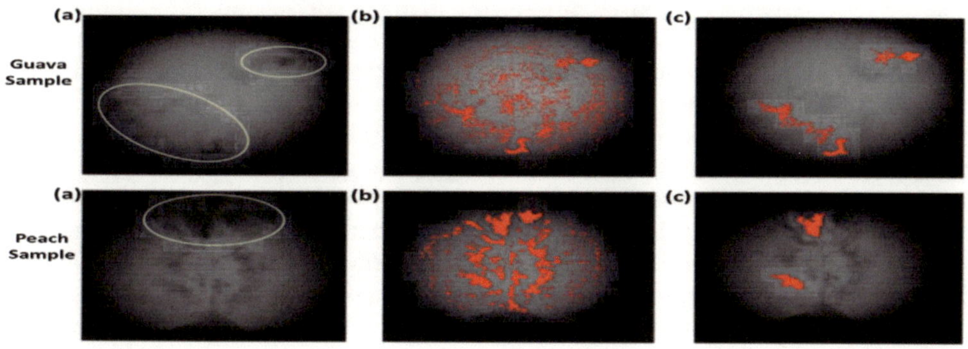

Figure 15.
Illustrates effect of morphological filtering: (a) X-ray image of a sample, (b) segmented spots after adaptive thresholding of (a), and (c) morphological filtering of (b) with three iterations.

affected by the selection of sub-image and it decreases as the sub-image size increases. Where, the detection accuracy slightly increased to 95% (guava) and 98% (peach), by reducing the sub-image size to 12 $\times$ 12 in the adaptive segmentation procedure. Furthermore, there are many studies on the applications of X-ray imaging techniques for inspection and evaluation quality that have been reported in the field of food and agriculture.

For example, [57] used the X-ray image to create a method of measuring the mass of wheat grains via calculating the total grey value. Detecting internal defects in grains or seeds by applying X-ray imaging has also shown promising results where, [58, 59] proved that the X-ray imaging technique can identified wheat grains infested by weevils. Also, the apple bruises were detectable using X-ray imaging and the extracted image features can be used to sort defective apples [60]. As well, [61] concluded that digital X-ray images can detect the internal disorder that leads to tissue breakdown such as the watercore in apples.

4.3 Ultraviolet (UV) imaging

Ultraviolet (UV) radiation has a shorter wavelength and higher energy than visible light band covers the wavelength range 100–400 nm. Moreover, UV radiation was divided into three bands UVA (320–400 nm), UVB (290–320 nm), and UVC (100–290 nm) is the most damaging type of UV radiation. However, it is completely filtered by the atmosphere and does not reach the earth's surface. Indeed, [30] mentioned that the imaging cannot be used in the region below

290 nm while UVB scattered more than the UVA and visible light. In reflected-UV imaging, UV illumination reflects off a scene then recorded by a UV sensitive camera while in UV fluorescence imaging. Also, UV illumination stimulates fluorescence at a longer wavelength UVA than the UV excitation source. The resulting fluorescence is typically in the visible light band. Applications of the ultraviolet band are varied. They include lithography, industrial inspection, microscopy, lasers, biological imaging, and astronomical observations [3]. In addition, [30] mentioned that UV light tends to be absorbed strongly by many organic materials and makes it possible to visualize the surface topology of an object without the light penetrating the interior parts. In the UV imaging field, little research on the UV camera as the main part of computer vision system based on image processing system has also been carried by the researchers. For example, [62] pointed to that running research focused on the study of reflected ultraviolet imaging (UV) technique, its potential of detecting defects in mangoes, and to develop a computer vision system that could find the reflected area on injured or defected mango's surface. So, they studied the possibility of a reflected UV imaging technique for the detection of defects on the surface area of mango. concluded that the distinction between RGB color and reflected UV imaging is very clear as shown in **Figure 16**. The band-pass filter of 400 nm wavelengths was found more suitable to detect the defected or ruptured tissues of mangoes. It might be due to the high photographic value of the UV-A band and since the reflected UV photography well performed over 360 nm as mentioned by [63]. Accordingly, an algorithm for defect segmentation can be developed and CVS could combine with a UV camera and a software algorithm to detect injuries. In this context, [64] developed and tested a prototype UV-based imaging system for real-time detection and separation of dried figs contaminated with aflatoxins as shown in **Figure 17**. The prototype system was tested by using 400 dried figs.

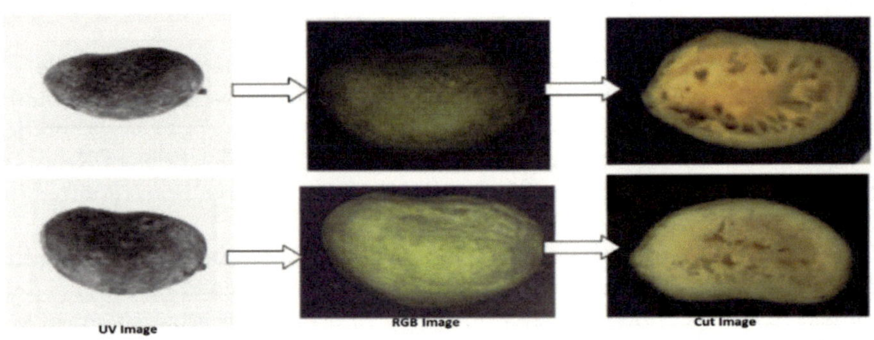

Figure 16.
Shows different images of shriveled mangoes.

In the final analysis, the prototype system achieved a 98% success rate in the detection and separation of the dried figs contaminated with aflatoxins. Also, [65] have built a simple computer vision system to detection of anthracnose infection and latex stain by using a low-cost webcam under UV-A illumination. The UV-fluorescence imaging technique has been selected for detecting areas on dried figs that are contaminated with aflatoxin [66]. Similarly, freeze-damaged oranges were also detected using the ultraviolet (UV) fluorescence method by [67] at 365 nm. Furthermore, [68] found that a UV-based computer vision system was effective in identifying stem end injuries in citrus fruits, which was used for fruit sorting. Likewise, [69] introduced a modern method based on UV imaging technique and processing images under a 365 nm UV light for separating pistachio nuts

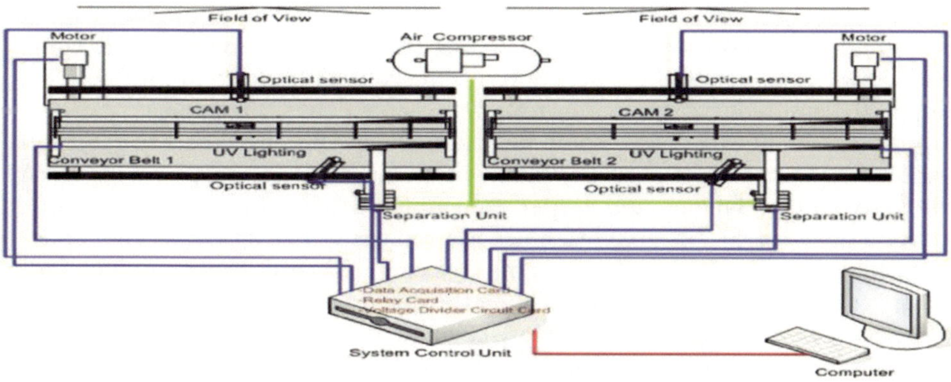

Figure 17.
Schematic of UV-based imaging system for aflatoxin contamination detection.

contaminated with aflatoxins. Accordingly, [70] indicated that various hidden defects inside fruits and vegetables can't be recognized by conventional systems, in contrast, can be identified by the reflected UV imaging technique. In this regard [71] mentioned that there is an important band (i.e., 365 nm) was identified during UV band selection in the application of UV-fluorescence imaging technique for inspecting aflatoxin contamination. Also, for more than 30 years [72] used UV photographs for aflatoxin-producing molds that were identified as gray or black colonies, whereas molds not producing aflatoxins appeared as white colonies.

4.4 Visible light imaging

The visible light spectrum is defined as the segment of the electromagnetic spectrum that the human eye can view. This visible light band is located in between ultraviolet (UV) and infrared (IR) regions, whose wavelength ranges from 400 to 700 nm as shown in **Figure 18**. Visible light is partly absorbed or diffused and partly reflected from the surfaces of objects, giving them the color, we perceive. This visible light region consists of red, orange, yellow, green, blue, and violet waves. Obviously, each color wave is defined with a specific wavelength where violet-blue is in the area of from 400 to 475 nm, the yellow-green color of about average 550 nm, and red is located in the area of 700 nm [28, 73]. Additionally, [74] mentioned that when the light falls on an object, it is usually reflected, absorbed, or transmitted. The intensity of these phenomena depends on the nature of the material and that specific wavelength region of the electromagnetic spectrum that is being used.

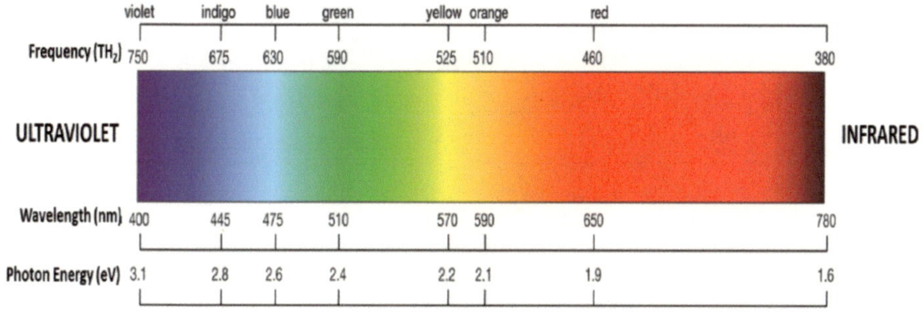

Figure 18.
Shows the characteristics of visible light on the electromagnetic spectrum.

The visual band of the electromagnetic spectrum is the most familiar in all our life activities, it is not surprising that imaging systems based on this visible light band outweigh by far all the others in terms of scope of application [3]. More simply, the emitted, transmitted or reflected visible light from an object carries information about that object which facilitates the quality inspectors to get information concerning the quality. So, visible-light imaging systems, play a significant role to see clearer, farther, and deeper and gaining detailed information about different objects [75]. Imaging machines base on the visible light spectrum or as called color imaging systems has become an extremely significant technique for nondestructively inspecting and assessing the quality of agricultural and food products. So, the color imaging machines are considering a promising technique currently applied for quality measurement of fresh and processed food. Visible light imaging machines operation is summarized, by acquiring images under illumination standard conditions, pixels processing, and analyzing the whole image which can classify and quantify objects. Also, visible light machine vision systems scan and sort millions of items per minute and provide fast, objective, robust measurement, and detailed characterization of color uniformity at a pixel-based level [76–78]. The simplest machine vision system is mainly composed of a lighting system attached with a camera, and a computer equipped with an image acquisition board as shown in **Figure 19**, [79, 80]. The accuracy, speed, and consistency of these technological developments represented in visible light imaging have greatly increased their applications in multiple fields in agriculture and food such as applications of pre- and post-harvest, food industry, baking industry, cereals, meat, fish, poultry, fruits and vegetable industry, and liquids. For instance, some agricultural research has been focusing on using machine vision systems based on color imaging and developing algorithms to count agricultural elements, mainly vegetables, and fruits to determine the full maturity, production, and harvest dates [81, 82]. In this regard, [83] presented a method for identifying and counting fruits from images acquired in cluttered greenhouses. The results showed a strong correlation, 94.6%, between the automatic and manual counting data. As well, [84] estimated the mango crop yield using image analysis to count mango from orchard images, and that is through segment the pixels of the images into two groups, fruit, and background, utilizing color and texture information. Then, the mangos were identified to count the number of fruits in the image. The automatic results achieving a strong correlation of 0.91.

The number of green apples was determined by using RGB color images under natural illumination [85]. Similarly, image analysis was used before harvesting to

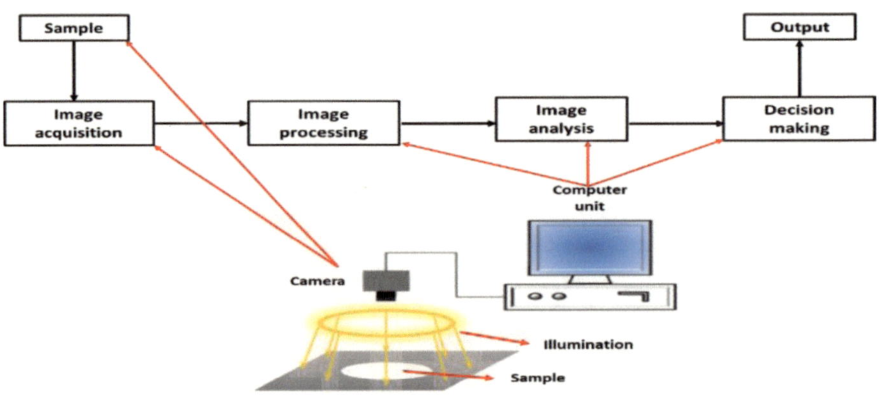

Figure 19.
Illustrates the basic components of the machine vision system.

counting the number of ripe and unripe fruits [86]. Also, [87] proposed a machine vision-based visible and NIR hyperspectral imaging method for automating yield estimation of golden delicious apples on trees at different growth stages. Subsequently, many successful attempts were recorded to use automatic vision systems based on image processing for quality control in post-harvest stages. In this context, [13] inspected potato tubers according to some sensitive quality features such as color, size, mass, firmness, and the texture homogeneity of potato surface **Figure 20** through a developed automated vision system. Concluded that the vision system can be applied as a non-destructive, precise, and symmetric technique in-line inspection. Additionally, [88] applied a computer vision system and machine learning algorithms to obtain a prediction model for cherry tomato volume and mass estimation and the results achieved an accuracy of 0.97. Also, the carrot was graded using a machine vision system and the results showed that the constructed image acquisition system success to extract the feature parameters of the carrot accurately [89]. As well, [6] sorted irregular potatoes using the RGB color imaging technique. Furthermore, ripeness determination of grape berries and seeds was performed using image analysis [90]. In a similar trend, the visual quality of agricultural grain is one of the extremely important issues in grain commercialization, which is assessed based on color, shape, and size, which generally impact the product's market price.

The main problems associated with the process of grain quality inspection are the high probability of error occurrence and the difficulty of standardizing the results. So, many proposals have been presented in the field of computer vision systems to assist visual inspection quality of several agricultural grains such as rice [91–94]. Also, about beans grains, many studies show the need and the importance of computer vision systems based on image processing for bean inspection [95–99]. In this context, [100] presents a machine vision system (MVS) for visual quality inspection of beans composed of a set of hardware consists of a board that includes an image acquisition chamber, a conveyor belt controlled by a servo motor, and a feeding mechanism and software for segmentation, classification, and defect detection as shown in **Figure 21**.

The results of offline experiments for segmentation, classification, and defect detection achieved, respectively, the average success rates of 99.6%, 99.6%, and 90.0%. While the results obtained in the online mode demonstrated the robustness and viability of this machine vision system, with average success rates of 98.5%, 97.8%, and 85.0%, respectively, to segment, classify, and detect defects in the grains

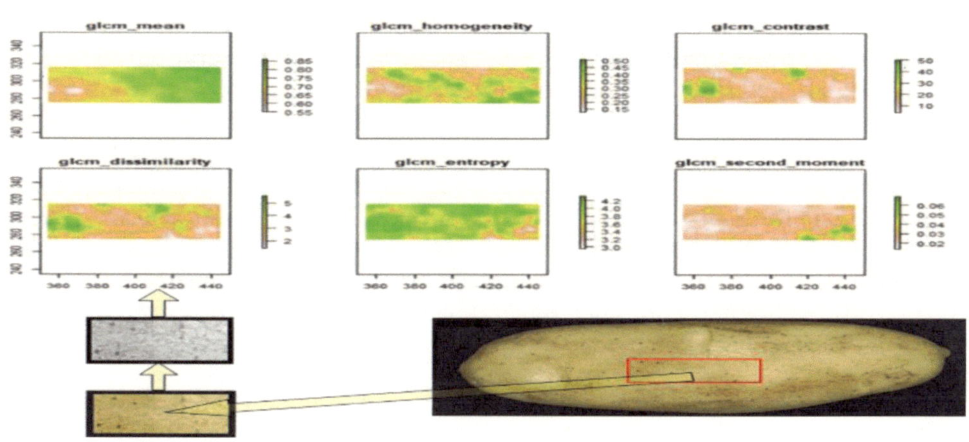

Figure 20.
Describe the extracting distinctive texture features of potato.

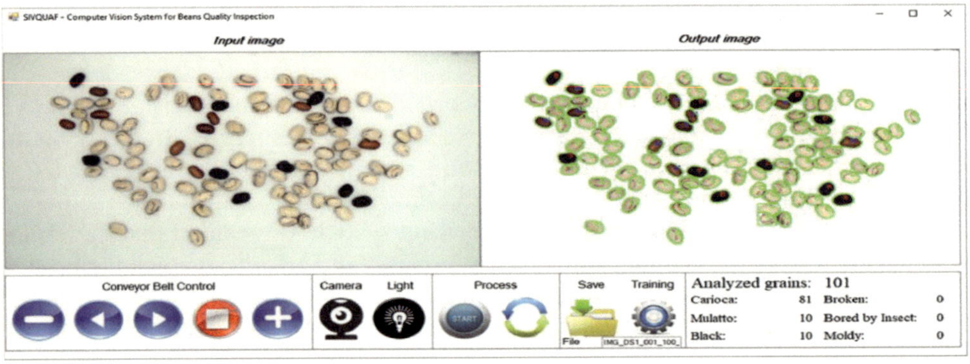

Figure 21.
Interface of the software of machine vision system for beans grain.

contained in each analyzed image. In the field of inspection quality of meat, imaging methods have been recently applied to visually assess meat and foodstuff quality on the processing line based on color, shape, size, surface texture features [101, 102]. A machine vision system with a support vector machine was utilized to grade the beef fat color. The highest performance percentage of the SVM classifier obtained was 97.4% [103]. Moreover, [104] mentioned that RGB color imaging has been a promising technique for predicting the color of meat. The moisture content of cooked beef joints was correlated with its color, using an RGB color imaging system [105]. The combination of machine vision, linear and nonlinear classifiers was employed for the automatic sorting of chicken pieces like breast, leg, fillet, wing, and drumstick. The results revealed that the total accuracy of online sorting (highest speed about 0.2 $m.s^{-1}$) was 93% [106]. As well, in the case of fish, visible-light imaging technology has been able to successfully predict the breed, species, quality, and gender of the fish [102]. Also, a machine vision method was used to evaluate the freshness of some fish. The best classification performance was achieved by the support vector machine classifiers with an 86.3% accuracy rate in the assessment of the carp fish based on its freshness [107].

4.5 Infrared (IR) imaging

Infrared (IR) radiation is a type of electromagnetic spectrum, a continuum of frequencies produced when atoms absorb and then release energy that's invisible to human eyes but that we can feel as heat. IR radiation is emitted by any object with a temperature above absolute zero and the most common sources of infrared radiation are the sun and fire. IR radiation exists in the electromagnetic spectrum at frequencies above those of microwaves and exactly below those of the red visible light band, hence it was called "infrared" as shown in **Figure 22**. IR frequencies range from about 300 (GHz) up to about 400 (THz). Waves of infrared radiation are longer than those of visible light, ranging from 0.75 to 1000 μm, and are divided into near (NIR, 0.78–3 μm), Mid-Infrared (MIR, 3–50 μm), and Far-Infrared (FIR, 50–1000 μm) as defined by the International Organization for Standardization (ISO 20473, 2007) optics and photonics-spectral bands, [108, 109]. The infrared spectrum (IR) is invisible to the human eye but has a wide range of uses in modern technology.

Different wavelengths (NIR, MIR, and FIR) of IR radiation have many different applications. The sources of IR introducing great technological advancements in imaging, thermal imaging, motion detection, gas analyzing, monitoring, and environmental health analysis, etc. IR imaging is widely used in the military, medical,

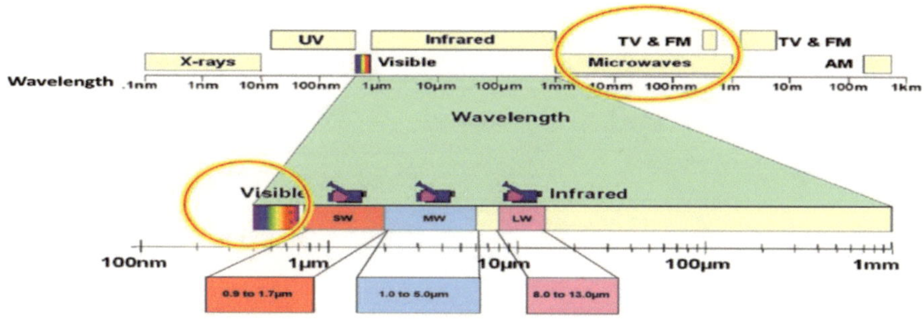

Figure 22.
An image of infrared wavelengths within the electromagnetic spectrum.

scientific, and industrial fields, since it is able to create a visual with an otherwise non-visible wavelength band to the human eye [110, 111]. Recently, multispectral, and hyperspectral imaging systems based on the IR spectrum, have been used for developing and evaluating most agricultural and food processing operations, such as tracking and estimating the quality of agricultural and food products.

4.5.1 Near-infrared (NIR) imaging

NIR techniques are used for qualitative analysis of agricultural and food products such as grain, fruit, vegetable, meat, fish, chicken, beverages, and dairy products. One of the most important of these techniques, and the most widespread is NIR imaging and spectroscopy, which offers a rapid, non-destructive, and cost-effective method. Development in instrumentation and data analysis techniques of NIR imaging and spectroscopy, expanded the application range to chemical analysis, agricultural and food product analysis, and more. So, NIR imaging is one of the preferred quality monitoring methods in the food industry [112, 113]. Conventional methods of agricultural and food product monitoring are time-consuming, expensive, and require sample destruction. So, the trend was towards fast, accurate, and non-destructive methods. NIR spectroscopy was established as a non-destructive method for quality analysis of food materials as mentioned by [114]. Ordinarily, when the IR radiation interacts with matter, the energy can be absorbed and result in molecular vibrations for example stretching, bending, rocking, wagging, and twisting. Hence, a change occurs in the electric dipole moment (change in the positive–negative charge separation) of the molecule, and the molecules transition to different vibrational levels as shown in **Figure 23**. These, transitioning from 1st to the 2nd, 3rd, or 4th excited state are known as overtones, and NIR spectroscopy measures these overtones [115]. So, NIR spectroscopy can therefore be used to study organic samples, which contain chemical bonds such as (C-H, O-H, N-H) because these functional groups absorb the energy from radiation in this region [116].

Therefore, [75, 117] mentioned that instead of individual compounds, major functional groups were assigned to specific NIR regions, where at a given wavelength range, a chemical bond will absorb the energy at a specific frequency when the energy matches the energy required to induce a vibrational response **Figure 24**.

Hyperspectral imaging based on the NIR band is the most widely used in the quality determination of agricultural and food products. However, NIR spectroscopy assessments do not contain spatial information, which is important to many food inspection applications. Furthermore, the inability of NIR spectrometers to capture internal constituent gradients within food products may lead to discrepancies between predicted and measured composition. Also, conventional Vis/NIR

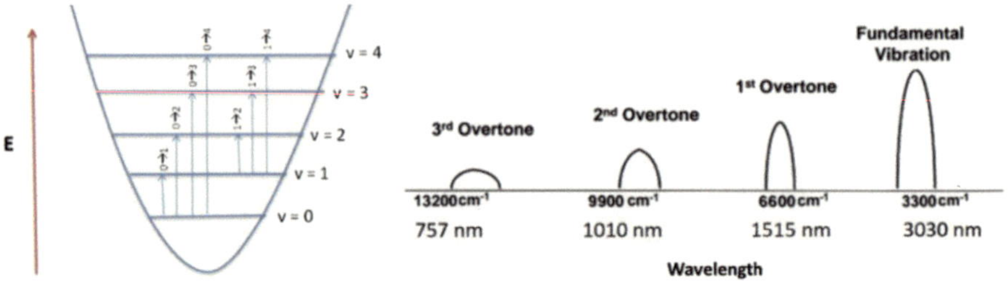

Figure 23.
Shows different vibrational levels for molecules and overtones transitions.

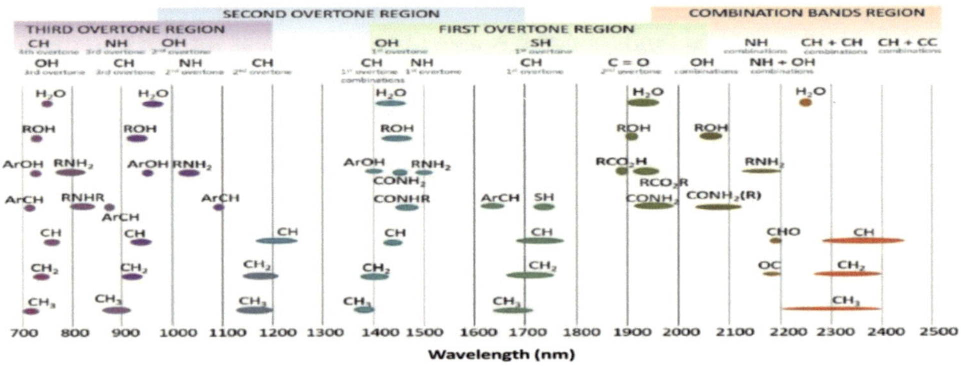

Figure 24.
Shows major analytical bands and relative peak positions for major NIR absorptions.

imaging provides only spatial information and does not supply any spectral information, which may lead to deficiencies in monitoring and evaluating the quality of products [118–120]. To overcome this, multispectral and hyperspectral imaging systems have been developed to combine images that contain spatial and spectral information, acquired at narrow wavebands, sensitive to features of interest on the object.

4.5.2 Hyperspectral imaging (HSI)

Hyperspectral imaging (HSI) or spectroscopic imaging is one of the most promising emerging technologies that integrates conventional imaging and spectroscopy to acquire both spatial and spectral information from an object. Although HSI was originally developed for remote sensing, it has recently emerged as a powerful process analytical tool for automatic non-destructive analysis of agricultural and food products [6, 121–125]. Where, the non-destructive, and flexible nature of HSI makes it an attractive process analytical technology for the identification of critical control parameters that impact finished product quality. As a result, expected [126, 127] that HSI will be increasingly adopted as a process analytical technology for quality monitoring of agricultural products and the food industry, as has already been the case in the pharmaceutical industry. There is an equally significant aspect, where the importance of the HSI system is that it consists of hundreds of neighboring wavebands for each spatial position (pixel) within the image. Hence, the spectrum considers like a fingerprint that can be used to characterize the composition of that pixel. HSI images are three-dimensional blocks of data, including two dimensions as spatial position and one spectral dimension, so this HSI is known as

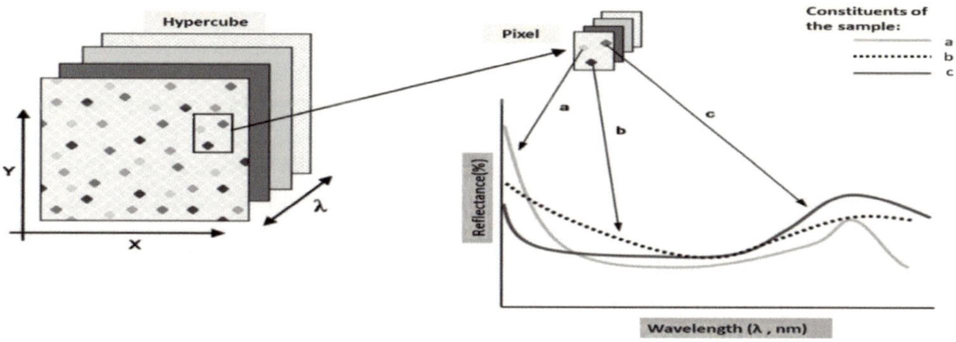

Figure 25.
Schematic of HSI hypercube, the spectral and spatial dimensions relationship.

hypercubes, as clarified in **Figure 25**. Each hypercube consists of 50–300 images acquired at different wavelengths with a spectral resolution of 1–10 nm. Another significant factor is that the hypercubes (HSI) permit the visualization of biochemical constituents of a sample, as separated forms into areas of the hyper image [6, 122, 128–132]. In brief, the main idea of the HSI imaging system running is that when the electromagnetic spectrum beam incident on the sample during sample analysis, the radiation turns into forms of reflection, scattering, absorption, and emit electromagnetic energy obtaining different patterns in specific wavelengths, due to the difference in chemical composition and physical structure of the sample. As a consequence, each element has a spectral fingerprint declaring its chemical composition. So, differences in the chemical concentration of the constituents of the sample lead to different reflectance or absorbance values in some main wavelengths [130, 131, 133].

Generally, the structure of the HSI system consists of some major components: lens, spectrograph, camera, translation stage, illumination unit, and computer system **Figure 26**. Then, when the sample is highlighted by diffuse illumination such as tungsten-halogen or LED source. then, the sample reflects the light to the lens and is separated into its component wavelengths by diffraction optics contained in the spectrograph, then a two-dimensional image (spatial and spectral dimensions) is formed on the camera and saved on the computer system [122, 134].

Consequently, these technological developments in HSI techniques based on NIR as a measuring non-destructive method, accurate, reliable, and fast for quality and safety analysis, have greatly increased the applications in a wide range of

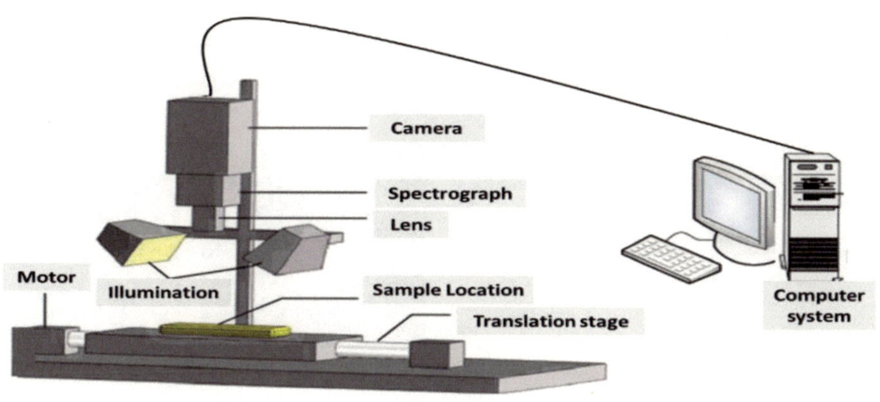

Figure 26.
Diagram of the hyperspectral imaging system components.

agricultural and food products. In this paragraph, some applications will be listed that demonstrate the capability of HSI in the field of food to perform classification, defect and disease detection, and assessment of some chemical characteristics. Furthermore, [135] clarified that the HSI systems can be used for the discrimination of different types of grains, including maize, wheat, barley, oat, soybean, and rice seed, etc. For instance, [136] developed indices for Norway spruce (Picea abies) seeds screening through applying HSI at different wavelengths 1310, 1710, and 1985 nm and the results showed a good classification, recommending the possibility to build inexpensive devices. As well, [137] used HSI based on the NIR band to explore the influence of grain shape and texture on the spectral variation represented in three kinds of cereal barley, wheat, and sorghum using PCA and gradients classification. Concluded that the results of classification gradient images and PC score plots were 91.18, 89.43, and 84.39% respectively, and all were influenced by kernel topography. An equally significant aspect is determining the viability of seeds by applying HSI at different spectral ranges (400–1000, and 1000–2500 nm). Visualization of treated and non-treated corn seeds was also achieved with HSI. The results demonstrated that the spectral range in the 1000–2500 nm performed better in exploring the seed viability [138]. Also, [139] classified viable and non-viable kernels of different cultivars of barley, wheat, and sorghum by using the NIR-HSI system. The results showed that NIR hyperspectral imaging is capable to identify viable and non-viable kernels of different cultivars. In a study for industrial baking of sponge cakes [140], the production process required various quality indicators to be measured continuously such as moisture content and sponge hardness. The existing techniques for performing these measures, randomly selected sponges are removed from the production line, and then samples are manually cut from each sponge by a destructive method to test as shown in **Figure 27A**. In contrast, the authors used the NIR-HSI system with a spectral range of 900–1700 nm as a non-destructive method to predict both moisture and hardness of cake **Figure 27B**. The results showed that the moisture and hardness prediction models when using a PLS-R model were 0.99 and 0.98. Accordingly, concluded that HSI is a valid method for predicting sponge cakes' moisture content and hardness. This study established a proof of concept for a new stand-off cake moisture and hardness monitoring system. Additionally, this HSI system would provide the added advantage to record every product in an HS image, which leads to detect variations in the production process. Also, HSI systems were applied for the ripeness monitoring of a large number of different fruit varieties [141–147]. Also, defects or blemishes detection such as bruising in fruit [147–155]. Recent studies on

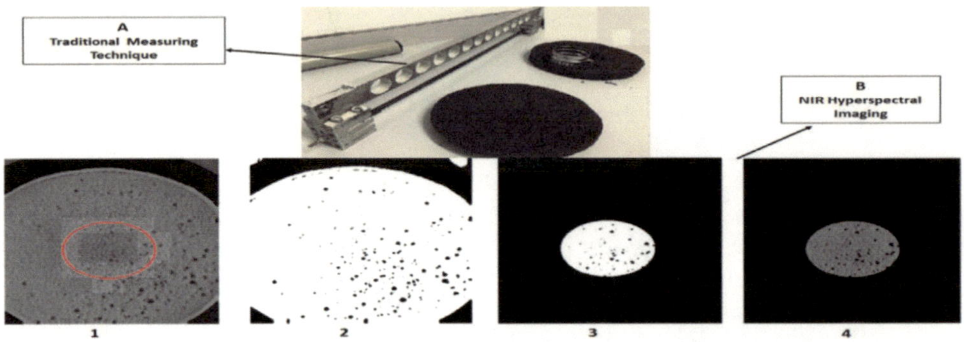

Figure 27.
Illustrated traditional measuring technique (A), and (B) NIR-HSI, (1) single band at 1450 nm, (2) binary image obtained indicating the location of the cavities (in black), (3) Binary mask selecting the center of the cake (white), air bubbles (black), and (4) cake image ready for spectral data extraction.

the safety inspection of agricultural products and livestock use multispectral imaging and HSI technologies. HSI methods have been used to determine the contamination of internal secretions on the surface of chickens, surface contamination in food processing, and fecal or foreign contamination of matter for apples and lettuce [118, 156, 157]. While studying the potential application of HSI for defect identification, apple and cucumber are two of the most popular food products that are being studied for bruises and frost injury defects, respectively.

Moreover, [158] took three varieties of apples to study the damage in apples and noted that the NIR region (700 and 900 nm) was more efficient at determining it. As a result, the NIR-HSI system from 900–1700 nm, to examine its application in the identification of bruises during various periods of storage after bruising was subsequently implemented by [159]. The spectrally reflective image analysis system has also been developed to assess defects on lettuce cut in the processing line. In particular, [160] algorithms have been identified to detect snails and worms. Another significant factor, where HSI systems proved not only to detect non-obvious bruises of fruits but also capable of assessing internal quality parameters such as soluble solid content, firmness, pH value, antioxidant, etc. [161–171].

4.5.3 Mid and far infrared imaging

Mid and far-infrared bands of EM radiation are an extremely useful part of the spectrum. Where, it can provide imaging in the dark, trace heat signatures, and provide sensitive detection of many biomolecular and chemical signals. However, the mid-infrared (MIR) band of the electromagnetic spectrum seems to contain valuable new information about some of the features needed to differentiate the samples, for example, the samples with diseases or some contamination or for quality inspection. Also, the recent development of light sources and imaging systems in MIR allows the use of multi/hyperspectral MIR imaging in many new applications as mentioned by [172, 173]. Also, signals of all IR radiation are known to be sensitive to leaf compounds such as water, lignin, and cellulose, which are essential to the functioning and structure of the leaf [173–175]. The thermal imaging technique is defined as a non-destructive, contactless, and rapid method for capturing the IR radiation from the object's surface. Where the surfaces of the hot objects emit electromagnetic waves in the IR region. Thermal imaging systems commonly capture radiation data from 7.5 μm up to 14 μm [176, 177]. This IR range is defined as the transmission window of the atmosphere characterized by the minimum attenuation of radiation [178]. Where the idea of a thermal imaging system based on captures temperature and spatial information simultaneously. Then, delivers the MIR data to be processed through a computer unit and provided in matrices forming called thermograms. From this point, there have been many successful attempts to apply thermal imaging systems as non-destructive and contactless methods to monitor the quality of many agricultural and food products. For example, but not limited, [37] developed an infrared thermal imaging system to detect infestation by Cryptolestes ferrugineus under the seed coat on the germ of the wheat kernels. Found that the overall classification accuracy for a quadratic function was 83.5% and 77.7% for infested and sound kernels, respectively, and for a linear function, it was 77.6% and 83.0% for infested and sound kernels, respectively, in pairwise discriminations. As well, [179] studied the feasibility of applying an IR thermal imaging system to classified fungal infections of stored wheat, the results prove that a thermal imaging system could be a useful tool to find if the wheat grain is infected by fungi or not, where the classification models gave a maximum accuracy of 100% for healthy samples and more than 97% and 96% for infected samples, respectively. Additionally, [180] developed a method to early detect apple bruising based on pulsed-phase thermography. The results indicated the high possibilities of the active thermography method for detecting defects up to several millimeters. Also, [181] conducted a follow-up study in which hyperspectral cameras were used equipped with sensors working in the visible and NIR (400–1000 nm), short-wavelength (1000–2500 nm), and thermal imaging camera in the MIR range (3500–5000 nm) to producing visualizations of bruises and providing information about bruise depth. the results obtained confirmed that the broad-spectrum range (400–5000 nm) of fruit surface imaging can improve the detection of early bruises with varying depths. Likewise, [157] adjusted an infrared lock-in thermography technique for the detection of early bruises on pears, the thermal emission signals from pears were measured using a highly sensitive MIR thermal camera. Found that the phase information of thermal emission from pears provides good metrics to identify quantitative information about both the size and the depth of damage for pears. In the same context, [182] developed a pulsed thermographic imaging system and explore its feasibility in non-destructively detecting bruised blueberries. The results demonstrated the feasibility of pulsed thermography to discriminate between bruised and healthy blueberries. Most recently, in the food processing sector, a study conducted by [183] indicated the possibility of monitoring and evaluating ovens systems through MIR imaging. Where this study aims to demonstrate the applicability of thermal imaging with image processing for the real-time evaluation of oven systems. A thermal camera was adapted to two different oven systems: a standard electric deck oven and a novel gas-fired baking oven with integrated volumetric ceramic burners as shown in **Figure 28**.

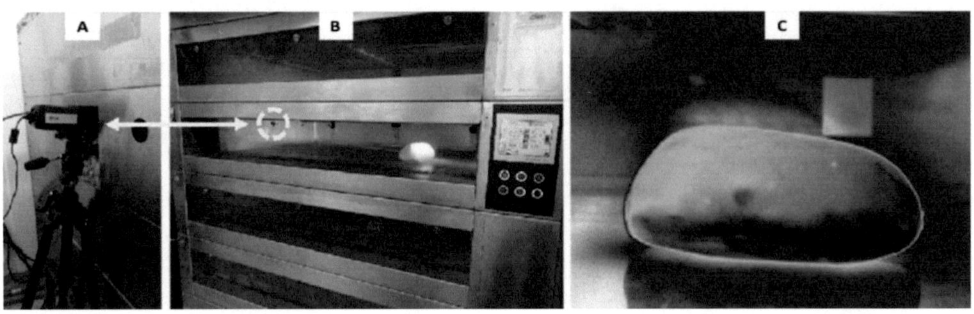

Figure 28.
Shows MIR system for monitoring the baking process: A) MIR camera, B) electric deck oven, and C) a thermogram captured during baking.

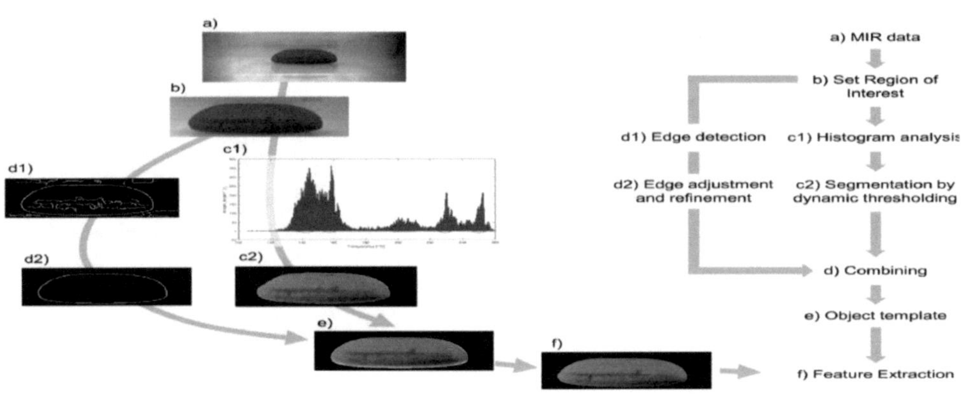

Figure 29.
Image processing steps with major operations.

MIR data with image processing are used to accomplish a time-resolved and automated monitoring of the baking process for oven system evaluation. Therefore, items to be baked were captured by the thermal camera, detected and feature extraction was performed to calculate the quality feature relevant such as texture homogeneity, temperature distribution, spatial dimensions (width and height), and the corresponding growth kinetic as shown in **Figure 29**. The results of the proposed study proved its fundamental qualification for comparing, monitoring, and evaluating different oven systems. In the final analysis, concluded that thermal imaging is an emerging and promising technique for the food industry and offers promising possibilities for inline process sensing and monitoring in the food sector.

4.6 Microwaves imaging (MWI)

Microwave radiation appears on electromagnetic radiation, between IR and radio waves. Where, microwaves refer to alternating current signals in the frequency range from 300 MHz to 300 GHz and (3 x 108 m/sec)/frequency, which gives you a wavelength range from 1 mm to 1 meter. These dimensions allow penetrating deep inside many optically not transparent mediums such as biological tissues, concrete, soil, wood, etc. In this regard, [3] indicated that radar is the dominant application of microwave imaging techniques. Because the imaging radar technique in the microwave band can collect data over any region at any time, regardless of the weather or ambient lighting conditions. Some radar waves can penetrate clouds, and can also see-through vegetation, ice, and extremely dry sand under non-standard conditions. The imaging radar works like a flash camera that provides microwave pulses to illuminate the target area and take a snapshot image. Where, imaging radar uses an antenna instead of a camera lens, attached with digital computer processing to record its images. In a radar image, one can see only the microwave energy that was reflected toward the radar antenna. There are many similarities between optical imaging, using a digital camera, and microwave imaging, using an antenna array as highlighted in **Figure 30**. In this type of imaging known as microwave holography, one or more antennas in the array illuminate the scene with a radiofrequency (RF) signal. Part of this signal is reflected in the other antennas, which record both the amplitude and phase of the reflected signal. These reflected RF signals are then processed to form an image of the scene [184–187]. Microwave imaging techniques have shown excellent capabilities in various fields such as civil engineering, biomedical diagnostics, safety, industrial applications, and have in the latest decades experienced strong growth as a research topic in the agricultural and food fields.

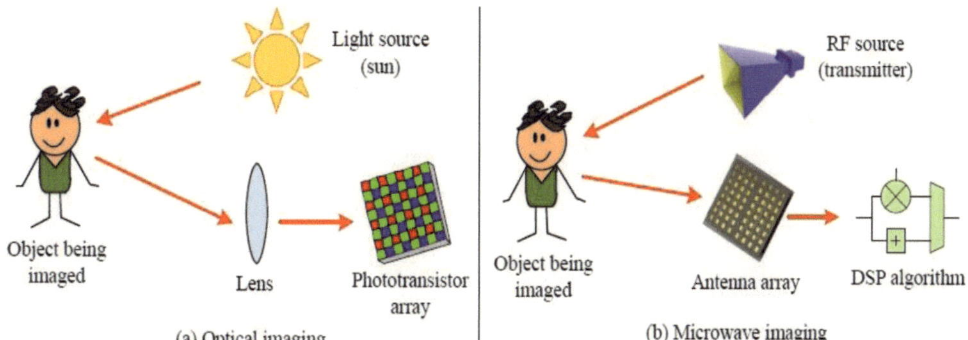

Figure 30.
Highlights similarities between visible light imaging, and microwave imaging.

Microwave imaging technology means the initial rapid screening of the hidden objects in an object's internal structure employing electromagnetic fields at microwave frequencies (300 MHz-30 GHz). Microwave images are maps of the electrical property distributions in dielectric samples [188, 189]. Therefore, microwave imaging for agricultural and food applications is nowadays of great interest, having the potential of providing information about the internal quality of agricultural and food products. There are three main reasons for the growing interest and rapid development of microwave-based methodologies, starting with the idea that the microwave band can penetrate all materials (unless ideal conductors), and the related scattered fields are representative of the overall volume of the object under test and not only of its surface; the second main interest reason that the microwave imaging modalities are very sensitive to the water content of the specimen, which makes them extremely suitable by particularly for food processing techniques; and thirdly it contactless concerning the specimen. Microwave imaging (radar tomography) has been used to evaluate the physical properties of food. In particular, the microwave imaging technique, able to identify the composition and the shape of biological materials, for the quality control of packed foods, and identify the degree of ripeness of fruits [190–192]. It could also be said through several investigations focused on the use of microwave technologies that microwave imaging techniques are used to probe inaccessible domains and to reveal the dielectric properties of the media that they penetrate. This technique aims to fully characterize the area in terms of positions, shapes, and complex permittivity profiles of the dielectric discontinuities (i.e., the scatterers). This aim is achieved by using inverse scattering algorithms to be analyzing the scattered field reflected by the material under consideration. Therefore, inverse scattering methods have been applied in many applications such as medical diagnosis, subsurface monitoring or geophysical inspection, and nondestructive evaluation and testing in various fields [193–198]. Accordingly, [199] focused on the application of microwave imaging technology for food contamination monitoring where the mechanism of this technology is based on transmission across the food sample to exploit the local dielectric proprieties variation that means a foreign object detection. Ordinarily, the microwave imaging system is composed of two main components hardware and software **Figure 31**, where the hardware part collects data, and the software process them to generate the output. Where the transmitter antenna generates EM waves toward the sample, that in food ambient it can be reasonably considered a homogeneous material, and a receiver antenna collects them. After their acquisition, dedicated software processes the data to generate the outputs where the detected intrusion is reported.

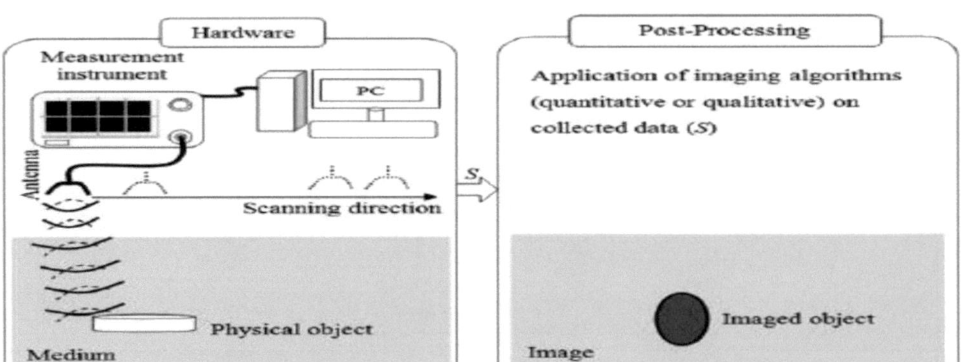

Figure 31.
Illustrates components of antenna microwave imaging system.

By the same token, [200] investigated the dielectric properties of fresh eggs during storage through frequency range 20–1800 MHz using an open-ended coaxial probe on thick albumen and yolk of eggs after 1–15 days of storage at room temperature. Also, [201] concluded that the dielectric properties of egg albumen and yolk were distinguished over the frequency range of 10–1800 MHz. Also, [202] presented a form of food security sensing using a waveguide antenna microwave imaging system to identify the health status of eggs. Therefore, proposed a waveguide antenna system with a frequency range of 7–13 GHz and a maximum gain of 17.37 dBi, with a scanning area of 30x30 cm^2. The results found that the proposed waveguide antenna microwave imaging sensing system could effectively identify the health status of many eggs very quickly. As a consequence, concluded that the waveguide antenna microwave imaging sensing system provides a simple, non-destructive, effective, and rapid method for food security applications. Images are undoubtedly the optimum technique in representing concepts to the human brain. Regardless of whether the product is fresh fruits or prepared food, color and moisture content are important attributes that food and agricultural engineers regularly look for. Therefore, [203] suggested an investigation focused on image acquisition technologies that can reveal the information of interest in 2-D using the visible, and non-visible (radar tomography) bands of radiation. The visible band was applied for color grading of oil palms and the computerized radar tomography was used to map the moisture content in grain. The results of this study found that the vision system correctly classified 92% of oil palms by four-color categories, and the radar tomography at 1 GHz frequency accurately mapped the homogeneity and heterogeneity in moisture content of grain over the moisture range 12–39%. At the same time, the microwave imaging technique is particularly useful for monitoring foods also after the packaging, without the necessity of opening the package. This is due to the microwave's ability to easily penetrate any type of non-metallic packages. Furthermore, it has the ability to identify unwanted or extraneous objects (such as glass or plastics pieces) embedded in food that cannot be detected with standard metal detectors. Microwave imaging technique in the frequency band from 8–12 GHz has been used to assess the contents of a package of cookies. The main purpose of applying this technique was to assess and ensure whether all the cookies within the package and ensure if their shape is preserved after the distribution or not. Through **Figure 32a**, the results concluded that the imaging technique in the microwave radiation range is capable of reconstruction of a package of cookies, and it can be noticed from the microwave image that one cookie is missed, and another is broken. Also, in a study to show the potentialities and abilities of microwave

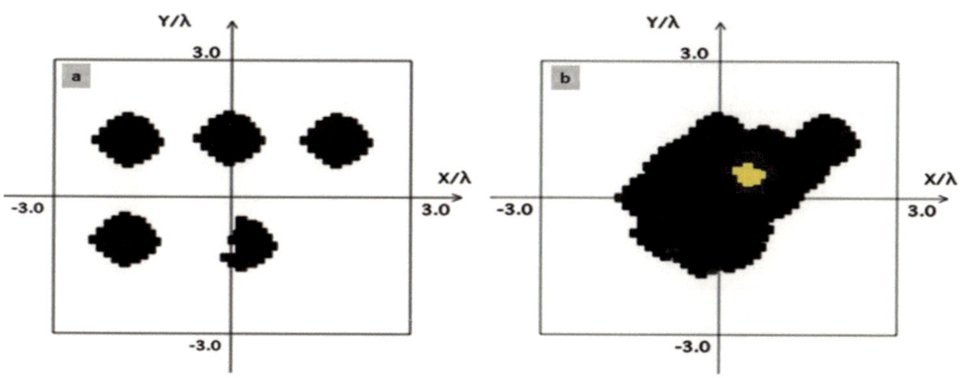

Figure 32.
Highlights microwave images, a) reconstruction of a package of cookies and b) identification of an abnormal object inside a piece of cheese.

imaging techniques for food processing applications. A sample of cheese was corrupted by placing a small piece of plastic material inside it to verify the ability of microwave imaging technology to detect this sample in a non-destructive manner. In this inspection technique, **Figure 32b** clearly shows the reconstruction of the dielectric distribution of the cheese piece through a microwave image, which clearly showed the presence of a small piece of plastic material and identified it as a yellow area [191].

4.7 Radio waves imaging

Radio waves are a type of electromagnetic radiation best-known for their use in communication technologies, such as television, mobile phones, and radios. According to NASA, radio waves have the longest wavelengths (1 mm to more than 100 km), also have the lowest frequencies, from about 3 kHz, up to about 300 GHz in the EM spectrum. The National Telecommunications and Information Administration generally divides the radio spectrum into nine bands **Table 1**. Low to medium frequencies, the lowest of all radio frequencies, have a long-range and are useful in penetrating water and rock. While, the high, very high, and ultra-high bands of radio frequencies include FM radio, broadcast television sound, public service radio, cellphones, and global positioning system (GPS). Moreover, super, and extremely high frequencies perform the highest frequencies in the radio band and are sometimes considered to be part of the microwave band. Imaging in the radio band, as in the case of imaging at the other end of the electromagnetic wave (gamma rays). Medicine and astronomy are the major applications of imaging in the radio band. Magnetic resonance imaging (MRI), or nuclear magnetic resonance scanner (NMR), is mostly known as a magnetic resonance imaging device. Because of its strong magnetism, the efficient polarization and further excites the focused proton singly included in water molecules present in the tissue. The technique of magnetic resonance imaging (MRI) is based on the magnetic field and pulses of radio radiation energy to evaluate the properties of objects, mostly applied for the diagnosis of various ailments internal to human and animal bodies. The main idea of the MRI technique is based on the magnetization of the atomic nuclei of the object using strong magnets and the nuclei rotate the magnetic field at variable speeds, which can be detected by the scanner and converted into usable data through Fourier transform. In the MRI technique, the hydrogen atom is used as a base atom because water is plentiful in all biological systems [3, 191, 204, 205].

Numbers	Radio bands	Frequency range	Wavelength range
1	Extremely Low Frequency (ELF)	<3 kHz	>100 km
2	Very Low Frequency (VLF)	3 to 30 kHz	10 to 100 km
3	Low Frequency (LF)	30 to 300 kHz	1 m to 10 km
4	Medium Frequency (MF)	300 kHz to 3 MHz	100 m to 1 km
5	High Frequency (HF)	3 to 30 MHz	10 to 100 m
6	Very High Frequency (VHF)	30 to 300 MHz	1 to 10 m
7	Ultra-High Frequency (UHF)	300 MHz to 3 GHz	10 cm to 1 m
8	Super High Frequency (SHF)	3 to 30 GHz	1 to 1 cm
9	Extremely High Frequency (EHF)	30 to 300 GHz	1 mm to 1 cm

Table 1.
Illustrates the nine bands classified of the radio spectrum.

Noting the magnetic nature of this MRI [204, 206] has mentioned that the low magnetic nature of the hydrogen protons which have different behaviors depending on the type of the tissues (e.g., lipids and water). The inspected object is placed within the magnet usually having 0.2–3.0 Tesla magnetic field power (T). This constant magnetic field is produced by radio-frequency pulses on the appropriate resonant frequency known as the Larmor frequency. It causes an excited state for the protons in the sample due to energy absorption. These protons generate radio waves, the emission can be detected by the receiver coil, producing an NMR signal. The basis for MR imaging is measuring the intensity signal of MR, accurate spatial placement of signal intensities, and cross-sectional representation of the signal intensities with the greyscale. Having high moisture content, agricultural products yield strong signals when applying an MRI technique. Therefore, [207] monitor the ripening of mangoes by using MRI technique and found that signal magnetic resonance intensity of the pericarp in MR images varied with the ripening stage. Also, [208] studied the prediction of sensory texture quality attributes of raw and cooked potatoes by NMR-imaging. MRI analysis on the obtained data and subsequent sensory analysis of the cooked potatoes displayed the high potential of employing advanced image analysis on MR-imaging data from raw potatoes to predict sensory attributes related to the texture of cooked potatoes. In short, concluded that MR-imaging besides giving well-known information about water distribution also gives information about anatomic structures within raw potatoes, which are considered important for the perceived textural properties of the cooked potatoes. Moreover, [209] designed an MRI apparatus characterized by its small, lightweight, and usable in an ordinary research room was devised for developmental research and quality estimation of foods and agricultural products. The proton-specified MRI was easy to operate and provided well-depicted images of internal structures, the distribution and mobility of water and oils, and susceptibility differences inside materials, demonstrating that the devised machine is useful for food and agricultural research. As well, [210] examined the changes in kiwi fruit tissue structure to evaluate the effect of storage conditions and found water migration in the direction of the outer region in the pericarp during storage. Also, [211, 212] applied MRI techniques to inspect the physicochemical changes of cherry tomatoes and found the potential of MRI techniques for tomato classification according to maturity. MRI is a technique that permits watercore detection without destroying the sample. From this point, [213] investigated the watercore distribution inside apple fruit (block or radial), and its incidence (% of tissue) by the non-invasive and non-destructive technique of MRI to obtain 20 inner tomography slices from each fruit and analyze the damaged areas using an interactive 3D segmentation method as shown in **Figure 33**. Apples with block watercore were grouped in Euler numbers between −400 and 400 with

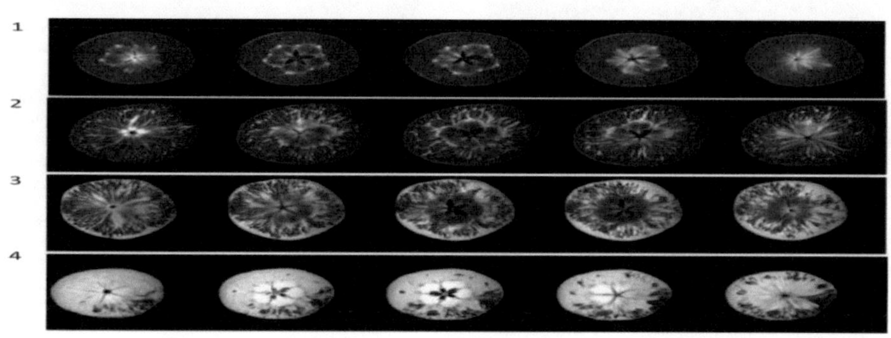

Figure 33.
Shows the MR images of central apple slices belonging to the four watercore levels, classified by three experts. (1) Sound apple; (2) light watercore; (3) medium watercore; (4) strong water core.

a small evolution. For apples with radial development, the Euler number was highly negative, up to −1439. Significant differences were also found regarding sugar composition, with higher fructose and total sugar contents in apples from the upper canopy, compared to those in the lower canopy location. Also, noted significantly higher sorbitol and lower sucrose and fructose contents were found in watercore-affected tissue compared to the healthy tissue of affected apples and compared to healthy apples. Additionally, [214] mentioned that by using MRI, the results of additional tests such as chemical analysis, oil and moisture distribution, sugar level, pH and physical analysis of structure, voids, the thickness of filling and coating, are immediately tested within seconds on the production line. Thus, the idea about the value of the MRI technique and its application in the food industry is going to improve and maintain the quality in processing, testing, and optimizing the parameters.

However, the high cost of imaging facilities is another barrier to the exploitation of MRI in the food industry. As well, [215] presented a detailed discussion on the fundamentals of MRI in the study of food materials. Also, [216] pointed to that the MRI is done with the use of an NMR instrument equipped with magnetic gradient coils. Where these coils have the capability to collect data spatially and create two-dimensional and three-dimensional images displaying diverse physicochemical characteristics. Likewise, [217] adopted the idea that fast and non-destructive solutions for sensing watercore would be readily accepted in the postharvest industry. Therefore, conducted a comparative study between X-ray CT and MRI as potential imaging technologies for detecting watercore disorder of different apple cultivars. After the acquisition of X-ray and MR images the 3D datasets of X-ray CT and MRI were matched, the images obtained on quantitatively identical fruit were compared. The results indicated that both MRI and CT were able to detect watercore disorder of different apple cultivars, however, the contrast in MRI images was superior as shown in **Figure 34**. Finally, concluded that the mean and variance of the frequency distribution of MRI and X-ray CT intensity appeared to be a parameter that allows the identification of healthy apples from affected fruit. A study by [218] provided a potential and detailed description of all components of the MRI system in agricultural fruits and vegetables for the assessment of maturity and quality parameters. As well, [219] used the MRI technique for non-invasive imaging of plant roots in different soils. Where used barley as a model plant to investigate the achievable image quality and the suitability for root phenotyping of six natural soil substrates of commonly occurring soil textures.

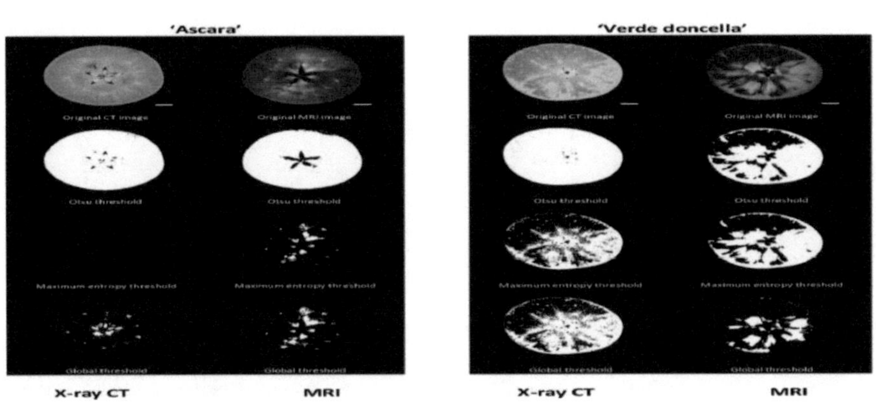

Figure 34.
Shows the X-ray CT (left) and MR images (right) cross-sections of sound Ascara and watercore Verde doncella fruit and their segmentation results.

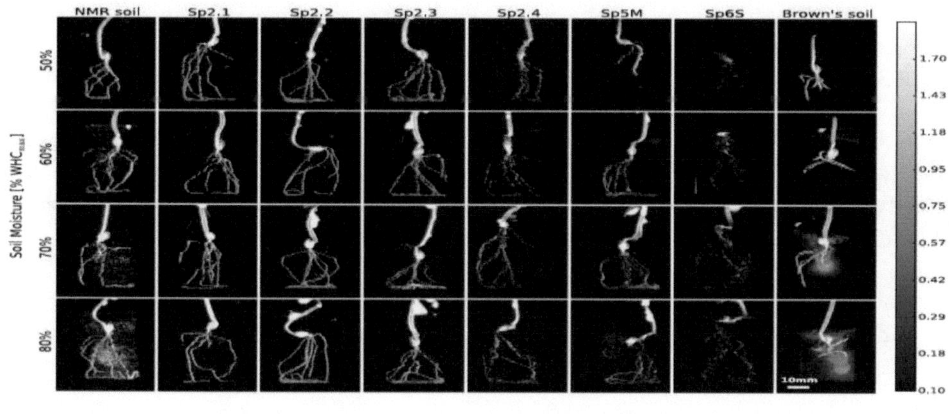

Figure 35.
MR Image for barley seedlings 3 days after sowing in eight different substrates at four different soil moisture levels.

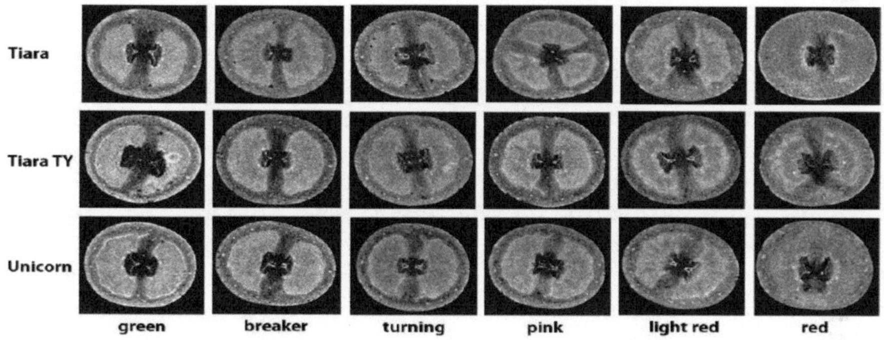

Figure 36.
Demonstrates the MR images of three cherry tomato cultivars Tiara, Tiara TY, and Unicornat, at different maturity stages.

The results are compared with two artificially composed substrates previously documented for MRI root imaging as shown in **Figure 35**. The results demonstrated that only one soil did not allow imaging of the roots with MRI. In the artificially composed substrates, soil moisture above 70% of the maximal water holding capacity (WHCmax) impeded root imaging. For the natural soil substrates, soil moisture did not affect MRI root image quality in the investigated range of 50–80% WHCmax. Concluded that with the characterization of different soils, investigations such as trait stability across substrates are now possible using non-invasive MRI. Subsequently, [220] presented a review conducted on the use of NMR/MRI techniques, for inspection of some agricultural fruits and vegetables, and explained the benefits of their implementation in the assessment of internal quality attributes such as internal defects, water content, nutrition content, maturity, fruit firmness, seed detection, physicochemical and microbiological quality in both commercial and industrial applications. Accordingly, concluded that the low-field nuclear magnetic resonance (LF-NMR) and MRI are viable technologies in assessing water status, which can significantly impact the quality of fruits and vegetables' texture, tenderness, and microstructure. Despite considerable developments in the quality measurement of fruits and vegetables and their products, the implementation of these techniques at an industrial level has been unsatisfactory. As well, [221] used MRI to study the changes in the internal structure of tomato fruit during development as a function of maturity. The internal structure of intact cherry tomato fruit at six different maturity stages (green, breaker, turning, pink, light red, and red) was measured using a series of two-dimensional (2D) MR images as shown in **Figure 36**. water content appears evenly distributed in the pericarp region from breaker to light red maturity stages.

MR signal intensity changes when different maturity stages are observed. Especially, signal intensity variation between the pericarp and locule regions is observed. Quantifying variations of signal intensity using a ratio of signal between pericarp and locule different regions enables the assignment of the maturity of cherry tomato. Additionally, concluded that since MRI provides detailed internal structure information, characterization of internal defects (e.g., bruises, voids, impact damage) and other quality factors is possible.

5. Conclusions

Inspecting and measuring the external and internal quality of agricultural and food products and assuring their safety from diseases and contamination, is one of the most important issues facing the food sector at present. This is a result of multiple and repeated complaints against agricultural producers and food manufacturers for the inability to meet quality requirements that meet the consumer's desires. When agricultural and food products do not meet quality standards and safety criteria, consumers lose faith in producers leading to the loss of these products' competitiveness in the market, and thus significant economic loss. With consumers rapidly growing demand for safer and better-quality food. So, agricultural producers and food manufacturers are working hard to eliminate sources of food contamination and achieve better quality. Although some systems are proposed to achieve food safety and quality by achieving a set of conditions that fall under the so-called good manufacturing practices (GMP) and hazard analysis and critical control point (HACCP) which represents the best way to achieve food security through all production steps. Unfortunately, with all these requirements for GMP and HACCP systems and others, they may not completely ensure the production of safe food free of contaminants and defects. So, it has become necessary to introduce modern technologies to quality inspect and detect blemishes and contamination and then reject these products that are not fit for human consumption. Therefore, the focus was on the development of non-destructive, modern, fast, reliable, and applicable methods that meet the needs of both food manufacturers and producers, as well as the desires of the consumer. So, the majority of all quality detection systems use electromagnetic wave measurements across all regions of the electromagnetic spectrum through imaging technologies. Gamma-ray and X-ray imaging technologies have high frequency and energy and are often used for irradiation, plant breeding applications, determining food quality, and food safety. Although applications of this technique are mainly used for research and development work, it has great potential to serve as a tool for the development of various plant varieties, assessment and quality assurance, and management practices for a wide range of agro-food practices. As well, UV and visible light imaging has proved itself to be very reliable and efficient for performing several tasks such as evaluating color, shape, size, and detect external defects. Additionally, these imaging systems can empower the agricultural and food industry with a new tool to detect defects and contaminations to ensure food safety and quality. Undoubtedly, with the evolution of a new generation of detectors and cameras, imaging within UV and visible light regions will have great potential in food defense and safety. Furthermore, IR and HSI systems can combine spectral and spatial data of a sample. For this reason, the HSI system became a standalone unit for non-destructive analysis of the physical, textural, and chemical parameters of the sample. So, the IR-HIS system has gained fame and a good reputation and is elaborately tested to predict chemical composition, detect defects, and adulterate agricultural and food products. Also, microwave and radio-wave imaging can improve the efficiency of real-time monitoring of food production, storing, and control quality chain. There appears to be an acceleration in the growth of hardware and software of imaging systems to overcome the limitations of this technology will help the agricultural and food industry in implementing the different imaging systems for rapid and in-line quality monitoring applications such as foreign material detection, discrimination external and internal quality attributes of agricultural and food products and detecting various defects and diseases. In conclusion, in this chapter, it has been shown that the different modern imaging technologies could provide unquestionably advantages for monitoring the quality and safety of agri-food production and processing.

Acknowledgements

The corresponding author thanks all members of this chapter for their scientific efforts. Also, extends his thanks to both Agricultural Engineering Research Institute (AEnRI), Agricultural Research Center (ARC); Menoufia University, Faculty of Agriculture, Department of Agriculture Engineering; Szent István University, Faculty of Agricultural and Environmental Sciences, Horticultural Institute; CREA (Council for Agricultural Research and Economics) of Treviglio, Italy; and King Fahd University of Petroleum and Minerals, Saudi Arabia, for their full support to the work team.

Conflict of interest

The authors declare no conflict of interest.

Author details

Ayman Eissa[1], Lajos Helyes[2], Elio Romano[3], Ahmed Albandary[4] and Ayman Ibrahim[5*]

1 Department of Agriculture Engineering, Faculty of Agriculture, Menoufia University, Egypt

2 Faculty of Agricultural and Environmental Sciences, Horticultural Institute, Szent István University, Gödöllő, Hungary

3 CREA (Council for Agricultural Research and Economics) of Treviglio (BG), Italy

4 Quality Assurance and Food Safety Unit, King Fahd University of Petroleum and Minerals, Saudi Arabia

5 Agricultural Research Center (ARC), Agricultural Engineering Research Institute (AEnRI), Egypt

*Address all correspondence to: aymanelgizawee@gmail.com

References

[1] Kingdom FAA. Mach bands explained by response normalization. Frontiers in Human Neuroscience. 2014; 8, 843. DOI:10.3389/fnhum.2014.00843

[2] Land EH, McCann JJ. Lightness and Retinex Theory. Journal of the Optical Society of America. 1971; 61: 1-11. DOI.org/10.1364/JOSA.61.000001

[3] Gonzalez RC, Woods, RE. Digital Image Processing. 2nd ed. Prentice Hall, Upper Saddle River; 2002. ISBN 10: 0201180758 / ISBN 13: 9780201180756.

[4] Cheremkhin PA, Lesnichii V V, Petrov NV. Use of spectral characteristics of DSLR cameras with Bayer filter sensors. Journal of Physics: Conference Series. 2014; 536: 012021. DOI:10.1088/1742-6596/536/1/012021.

[5] Bellman R. Adaptive Control Processes: A Guided Tour, Princeton University Press. 1961; DOI.org/10.1002/zamm.19620420718

[6] ElMasry G, Cubero S, Molto E, Blasco J. In-line sorting of irregular potatoes by using automated computer-based machine vision system. Journal of Food Engineering. 2012; 112: 60-68. DOI: 10.1016/j.jfoodeng.2012.03.027

[7] Ceamanos X, Valero S. Processing hyperspectral images. In: Nicolas B, Mehrez Z, editors. Handbook of Optical Remote Sensing of Land Surface. Elsevier; 2016. p. 163-200. DOI:10.1016/C2015-0-01220-5

[8] Yao-Ze F, Sun D-W. Application of Hyperspectral Imaging in Food Safety Inspection and Control: A Review. Critical Reviews in Food Science and Nutrition. 2012; 52: 1039-1058. DOI: 10.1080/10408398.2011.651542

[9] Kumar S, Kapil P, Bhardwaj Y, Acharya US, Gupta C. Anomaly Detection in Fruits using Hyper Spectral Images. International Journal of Trend in Scientific Research and Development. 2019; 3(4): 394-397. DOI: 10.31142/ijtsrd23753

[10] Pau G, Fuchs F, Sklyar O, Boutros M, Huber W. EBImage—an R package for image processing with applications to cellular phenotypes. Bioinformatics. 2010; 26(7): 979-981. DOI: 10.1093/bioinformatics/btq046

[11] Haralick RM, Shanmugam K, Dinstein I. Textural Features for Image Classification. IEEE Transaction on Systems, Man and Cybernetics. 1973; 3 (6): 610-621. DOI: 10.1109/TSMC.1973.4309314

[12] Park Y, Guldmann JM. Measuring continuous landscape patterns with Gray-Level Co-Occurrence Matrix (GLCM) indices: An alternative to patch metrics? Ecological Indicators. 2020; 109: 1-18. DOI: 10.1016/j.ecolind.2019.105802

[13] Ibrahima A, El-Bialeea N, Saad M, Romano E. Non-Destructive Quality Inspection of Potato Tubers Using Automated Vision System. International Journal on Advanced Science Engineering Information Technology. 2020; 10(6): 2419-2428. DOI: 10.18517/ijaseit.10.6.13079

[14] Romano E, Bergonzoli S, Pecorella I, Bisaglia C, De Vita P. Methodology for the definition of durum wheat yield homogeneous zones by using satellite spectral indices. Remote Sens. 2021;13: 2036. DOI.org/10.3390/rs13112036

[15] Yu H, Guo L, Kharbach M, Han W. Multi-Way Analysis Coupled with Near-Infrared Spectroscopy in Food Industry: Models and Applications. Foods. 2021; 10: 802. DOI.org/10.3390/foods10040802

[16] Gourvénec S, Tomasi G, Durville C, Di Crescenzo E, Saby C, Massart D, Bro R, Oppenheim G. CuBatch, a MATLAB® interface for n-mode data analysis. Chemometrics and Intelligent Laboratory Systems. 2005; 77 (1-2): 122-130. DOI: 10.1016/j. chemolab.2004.01.027

[17] Kolda TG, Bader BW. Algorithm 862: MATLAB tensor classes for fast algorithm prototyping. ACM Transactions on Mathematical Software. 2006; 32(4): 635-653. DOI: 10.1145/ 1186785.1186794

[18] Phan AH, Tichavský P, Cichocki A. Tensorbox: A Matlab Package for Tensor Decomposition. LABSP RIKEN: Saitama, Japan. 2012. http://www.bsp.b rain.riken.jp/~phan/tensorbox.php,

[19] Phan AH, Tichavsky P, Cichocki A. Low complexity damped Gauss–Newton algorithms for CANDECOMP/ PARAFAC. SIAM Journal on Matrix Analysis and Applications. 2013; 34: 126-147. DOI.org/10.1137/100808034

[20] Giordani P, Kiers HA, Del Ferraro MA. Three-way component analysis using the R package ThreeWay. Journal of Statistical Software. 2014; 57: 1-23. DOI: 10.18637/jss.v057.i07

[21] Helwig NE. Multiway: Component Models for Multi-Way Data. R Package Version 1.0-5. 2021; https://CRAN.Rprojec t. org/package=multiway. http://journal.a shspublications.org/content/123/5/898.

[22] Kossaifi J, Panagakis Y, Anandkumar A, Pantic M. Tensorly: Tensor learning in python. Journal of Machine Learning Research. 2019; 20: 1-6. https://creativecommons.org/lice nses/by/4.0/.

[23] Hao L, Liang S, Ye J, Xu Z. TensorD: A tensor decomposition library in TensorFlow. Neurocomputing. 2018; 318: 196-200. DOI: 10.1016/j. neucom.2018.08.055

[24] Gregoire P, Florian F, Oleg S, Michael B, Wolfgang H. EBImage - an R package for image processing with applications to cellular phenotypes. Bioinformatics. 2010; 26(7): 979-981. DOI: 10.1093/bioinformatics/btq046

[25] Grompone von Gioi R, Randall G. Unsupervised smooth contour detection. Image processing on line. 2016; 6: 233-267. DOI: 10.5201/ ipol.2016.175

[26] Grompone von Gioi R. A Contrario Line Segment Detection, Springer. 2014; DOI: 10.1007/978-1-4939-0575-1

[27] Otsu N. A threshold selection method from gray-level histograms. IEEE Transactions on Systems, Man, and Cybernetics. 1979; 9 (1): 62-66. DOI: 10.1109/TSMC.1979.4310076

[28] Starr C. Biology: Concepts and Applications. Thomson Brooks Cole; 6th edition. 2005; ISBN 0-534-46226-X.

[29] Williamson SJ, Cummins H. Light and Color in Nature and Art. American Journal of Physics. 1983; 1(2). DOI: 10.1119/1.13928

[30] Richards A. X rays and gamma rays: crookes tubes and nuclear light. In Alien Vision: exploring the electromagnetic spectrum with imaging technology. SPIE Press, Washington. 2001. ISBN-10 : 0819441422

[31] Saha GB. Physics and radiobiology of nuclear medicine. Springer, New York. 2006. DOI: 10.1007/978-0-387-36281-6

[32] Pires LF, de Macedo JR, de Souza MD, Bacchi OS, Reichardt K. Gamma-ray computed tomography to characterize soil surface sealing. Applied Radiation and Isotopes. 2002; 57: 375-380. DOI: 10.1016/S0969-8043(02)00098-2

[33] Naime JM, Cruvinel PE, Silva AM, Crestana S, Vaz CMP. Applications of X

and g-rays dedicated computerized tomography scanner in agriculture. In: Paulo, E.C., Luiz, A.C. (Eds.), Agricultural Tomography. Embrapa - CNPDIA, Sao Carlos. 2000; 96-104.

[34] Macedo A, Vaz, CMP, Naime JM, Jorge LAC, Crestana S, Cruvinel PE, Pereira JCD, Guimaraes MF, Ralisch R. Soil management impact and wood science recent contributions of Embrapa Agricultural Instrumentation Center using CT imaging. In: Paulo, E.C., Luiz, A.C. (Eds.), Agricultural Tomography. Embrapa- CNPDIA, Sao Carlos. 2000; 44-54.

[35] Tsukamoto T, Uchida H, Nakanishi H, Nishiyama S, Tsukada H, Matsuhashi S, Nishizawa NK, Mori S. $H_2{}^{15}O$ translocation in rice was enhanced by 10 μm 5-aminolevulinic acid as monitored by positron-emitting tracer imaging system (PETIS). Soil Science and Plant Nutrition. 2004; 50 (7):1085-1088. DOI: 10.1080/ 00380768.2004.10408578

[36] Kawachi N, Kikuchi K, Suzui N, Ishii S, Fujimaki S, Ishioka NS, Watabe H. Imaging of carbon translocation to fruit using carbon-11-labeled carbon dioxide and positron emission tomography. IEEE Transactions on Nuclear Science. 2011; 58(2): 395-399. DOI: 10.1109/ TNS.2011.2113192

[37] Manickavasagan A, Jayas DS, White, NDG. Thermal imaging to detect infestation by Cryptolestes ferrugineus inside wheat kernels. Journal of Stored Products Research. 2008; 44(2): 186-192. DOI: 10.1016/j. jspr.2007.10.006

[38] Rossel RAV, Taylor JJ, Mcbratney AB. Multivariate calibration of hyperspectral g-ray energy spectra for proximal soil sensing. European Journal of Soil Science. 2007; 58: 343-353. DOI: 10.1111/ j.1365-2389.2006.00859.x

[39] Tsukamoto T, Nakanishi H, Kiyomiya S, Watanbe S, Matsuhashi S, Nishizawa NK, Mori S. ^{52}Mn translocation in barley monitored using a positron-emitting tracer imaging system. Soil Science and Plant Nutrition. 2006; 52(6): 717-725. DOI.org/10.1111/ j.1747-0765.2006.00096.x

[40] Nakamura S, Suzui N, Nagasaka T, Komatsu F, Ishioka NS, Tanabata SI, Kawachi N, Rai H, Hattori H, Chino M, Fujimaki S. Application of glutathione to roots selectively inhibits cadmium transport from roots to shoots in oilseed rape. Journal of Experimental Botany. 2013; 64(4): 1073-1081. DOI: 10.1093/ jxb/ers388

[41] Fujimaki S, Suzui N, Ishioka NS, Kawachi N, Ito S, Chino M, Nakamura S. Tracing cadmium from culture to spikelet: noninvasive imaging and quantitative characterization of absorption, transport, and accumulation of cadmium in an intact rice plant. Plant Physiology. 2010; 152(4): 1796-1806. DOI: 10.1104/pp.109.151035

[42] Watanabe S, Iida Y, Suzui N, Katabuchi T, Ishii S, Kawachi N, Hanaoka H, Watanabe S, Matsuhashi S, Endo K, Ishioka NS. Production of no-carrier-added ^{64}Cu and applications to molecular imaging by PET and PETIS as a biomedical tracer. Journal of Radioanalytical and Nuclear Chemistry. 2009; 280: 199-205. DOI:10.1007/ S10967-008-7443-9

[43] Ishikawa S, Suzui N, Ito-Tanabata S, Ishii S, Igura M, Abe T, Kuramata M, Kawachi N, Fujimaki S. Real-time imaging and analysis of differences in cadmium dynamics in rice cultivars (Oryza sativa) using positron-emitting ^{107}Cd tracer. BMC Plant Biology. 2011; 11(1): 1-12. DOI: 10.1186/ 1471-2229-11-172

[44] Kiyomiya S, Nakanishi H, Uchida H, Nishiyama S, Tsukada H, Ishioka NS, Watanabe S, Osa A,

Mizuniwa C, Ito T, Matsuhashi S, Hashimoto S, Sekine T, Tsuji A, Mori S. Light activates $H_2{}^{15}O$ flow in rice: detailed monitoring using a positron-emitting tracer imaging system (PETIS). Physiol Plant. 2001; 113(3): 359-367. DOI: 10.1034/j.1399-3054.2001.1130309.x

[45] Mori S, Kiyomiya S, Nakanishi H, Ishioka NS, Watanabe S, Osa A, Matsuhashi S, Hashimoto S, Sekine T, Uchida H, Nishiyama S, Tsukada H, Tsuji A. Visualization of ^{15}O-water flow in tomato and rice in the light and dark using a positron-emitting tracer imaging system (PETIS). Soil Science and Plant Nutrition. 2000; 46(4): 975-979. DOI: 10.1080/00380768.2000.10409163

[46] Kotwaliwale N, Singh K, Kalne A, Jha SN, Seth N, Kar A. X-ray imaging methods for internal quality evaluation of agricultural produce. Journal of Food Science and Technology. 2014; 51(1): 1-15. DOI: 10.1007/s13197-011-0485-y

[47] Adedeji AA, Ngadi MO. Microstructural characterization of deep-fat fried breaded chicken nuggets using X-ray micro-computed tomography. Journal of Food Process Engineering. 2010; 34(6): 2205-2219. DOI: 10.1111/j.1745-4530.2009.00565.x

[48] Vavrik D, Jakubek J. Radiogram enhancement and linearization using the beam hardening correction method. Nuclear Instruments and Methods in Physics Research. 2009; 607(1): 212-214. DOI: 10.1016/j.nima.2009.03.156

[49] Ammari H. An Introduction to Mathematics of Emerging Biomedical Imaging. 1st ed; Springer-Verlag Berlin Heidelberg, Germany. 2008; X, 198p. DOI: 10.1007/978-3-540-79553-7

[50] Patton JA, Turkington TG. SPECT/CT physical principles and attenuation correction. Journal of Nuclear Medicine Technology. 2008; 36(1): 1-10. DOI: https://doi.org/10.2967/jnmt.107.046839

[51] Neethirajan S, Jayas DS, White NDG. Detection of sprouted wheat kernel using soft X-ray image analysis. Journal of Food Engineering. 2007; 81(3): 509-513. DOI: 10.1016/j.jfoodeng.2006.11.020

[52] Bianchini VJM, Mascarin GM, Silva LCAS, Arthur V, Carstensen JM, Boelt B, da Silva CB. Multispectral and X-ray images for characterization of Jatropha curcas L. seed quality. Plant Methods. 2021; 17(9): 1-13. DOI: 10.21203/rs.3.rs-28449/v1

[53] Einarsdottir H, Emerson MJ, Clemmensen LH, Scherer K, Willer K, Bech M, Larsen R, Ersbøll BK, Pfeiffer F. Novelty detection of foreign objects in food using multi-modal X-ray imaging. Food Control. 2016; 67: 39-47. DOI: 10.1016/j.foodcont.2016.02.023

[54] Al-Mezeini N, Manickavasagan A, Al-Yahyai R, Al-Wahaibi A, Al-Raeesi A, Khriji L. X-ray Imaging of Stored Dates to Detect Infestation by Saw-Toothed Beetles. International journal of fruit science. 2016; 16(1): 42-56. DOI: 10.1080/15538362.2015.1044692

[55] Nielsen MS, Lauridsen T, Christensen LB, Feidenhans'l R. X-ray darkfield imaging for detection of foreign bodies in food. Food Control. 2013; 30(2): 531-535. DOI: 10.1016/j.foodcont.2012.08.007

[56] Jiang JA, Chang HY, Wu KH, Ouyang CS, Yang MM, Yang EC, Chen TW, Lin TT. An adaptive image segmentation algorithm for X-ray quarantine inspection of selected fruits. Computers and electronics in agriculture. 2008; 60(2): 190-200. DOI: 10.1016/j.compag.2007.08.006

[57] Karunakaran C, Jayas DS, White, NDG. Mass determination of wheat kernels from X-ray images. American Society of Agricultural and Biological Engineers. Conference: 2004, DOI: 10.13031/2013.16742

[58] Haff RP, Slaughter DC. Real-time X-ray inspection of wheat for infestation by the granary weevil, Sitophilus Granarius (L.). Transactions of the ASAE. 2004; 47(2): 531-537. DOI: 10.13031/2013.16022

[59] Karunakaran C, Jayas DS, White, NDG. Soft X-ray inspection of wheat kernels infested by Sitophilus oryzae. Transactions of the ASAE. 2003; 46(3): 739-745. DOI: 10.13031/2013.13576

[60] Shahin MA, Tollner EW, Gitaitis RD, Sumner DR, Maw BW. Apple classification based on surface bruises using image processing and neural networks. Transactions of the ASAE. 2002; 45 (5): 1619-1627. DOI: 10.13031/2013.11047

[61] Kim S, Schatzki TF. Apple watercore sorting system using X-ray imagery. I. Algorithm development. Transactions of the ASAE. 2000; 43 (6): 1695-1702. DOI: 10.13031/2013.3070

[62] Patel KK, Kar A, Khan MA. Potential of reflected UV imaging technique for detection of defects on the surface area of mango. Journal of Food Science and Technology. 2019; 56(2): 1295-1301. DOI: 10.1007/s13197-019-03597-w

[63] Cosentino A. Practical notes on ultraviolet technical photography for art examination. Conservar Património. 2015; 21: 53-62. DOI: 10.14568/cp2015006

[64] Özlüoymak ÖB. Development of a UV-based Imaging System for Real-Time Detection and Separation of Dried Figs Contaminated with Aflatoxins. Journal of Agricultural Sciences. 2014; 20: 302-316. DOI: 10.15832/tbd.87873

[65] Nagle M, Intani K, Mahayothee B, Sardsud V, Muller J. Nondestructive mango quality assessment using image processing: inexpensive innovation for the fruit handling industry. In: Conference on international research on food security, natural resource management and rural development (Tropentag), Gottingen (Germany), 1-4. 2012.

[66] Özlüoymak ÖB. A Research on Separation System Design of Aflatoxin Contaminated Dried Fig [Ph.D.]. Adana, Turkey: Çukurova University; 2012.

[67] Slaughter DC, Obenland DM, Thompson JF, Arpaia ML, Margosan DA. Nondestructive freeze damage detection in oranges using machine vision and ultraviolet fluorescence. Postharvest Biology and Technology. 2008; 48(3): 341-346. DOI: 10.1016/j.postharvbio.2007.09.012

[68] Blasco J, Aleixos N, Gomez J, Molto E. Citrus sorting by identification of the most common defects using multispectral computer vision. Journal of Food Engineering. 2007; 83(3): 384-393. DOI: 10.1016/j.jfoodeng.2007.03.027

[69] Karami MA, Mirabolfathy M. Neural Network to Separate Aflatoxin Contaminated Pistachio Nuts. Acta Horticulturae. 2006; 726 (726): 605-610. DOI: 10.17660/ActaHortic.2006.726.103

[70] Kleynen O, Destain MF. Detection of defects on fruits by machine vision and unsupervised segmentation. In: AgEng conference, September 12-16. Technological Institute, Leuven, Belgium, 2004; 1006-1007. http://hdl.handle.net/2268/81214

[71] Steiner WE, Rieker RH, Battaglia R. Aflatoxin Contamination in Dried Figs: Distribution and Association with Fluorescence. Journal of Agricultural Food Chemistry. 1988; 36(1): 88-91. DOI: 10.1021/jf00079a022

[72] Yabe K, Ando Y, Ito M, Terakado N. Simple Method for Screening Aflatoxin-Producing Molds by UV Photography.

Applied and Environmental Microbiology. 1987; 53(2): 230-234. DOI: 10.1128/AEM.53.2.230-234.1987

[73] Robertson GL. Food packaging principles and practice, 3rd ed. CRC Press, Taylor and Francis group, UK. 2012. ISBN 9781439862414

[74] Jha SN. Non-destructive evaluation of food quality. Ist ed. Heidelberg, Springer; 2010. 18-22p. DOI: 10.1007/978-3-642-15796-7

[75] ElMasry G, Sun DW. Principles of hyperspectral imaging technology. In: Da-Wen Sun, editors. Handbook of Hyperspectral imaging for food quality analysis and control. 1st ed. Academic Press, San Diego, California, USA, 2010. 3-43p. DOI: 10.1016/B978-0-12-374753-2.10001-2.

[76] Lou W, Nakai S. Application of artificial neural networks for predicting the thermal inactivation of bacteria: a combined effect of temperature, pH and water activity. Food Research International. 2001; 34(47): 573-579. DOI: 10.1016/S0963-9969(01)00074-6

[77] Brosnan T, Sun DW. Inspection and grading of agricultural and food products by computer vision systems - a review. Computers and Electronics in Agriculture. 2002; 36 (2–3): 193-213. d oi.org/10.1016/S0168-1699(02)00101-1

[78] Wu D, Sun DW. Colour measurements by computer vision for food quality control-a review. Trends in Food Science and Technology. 2013; 29 (1): 5-20. DOI: 10.1016/j.tifs.2012.08.004

[79] Chen YR, Chao K, Kim MS. Machine vision technology for agricultural applications. Computers and Electronics in Agriculture. 2002; 36(2): 173-191. DOI: 10.1016/S0168-1699(02)00100-X

[80] Hong H, Yang X, You Z, Cheng F. Visual quality detection of aquatic products using machine vision. Aquacultural Engineering. 2014; 63: 62-71. DOI: 10.1016/j.aquaeng.2014.10.003

[81] Payne A, Walsh K, Subedi P, Jarvis D. Estimating mango crop yield using image analysis using fruit at 'stone hardening' stage and night time imaging. Computers and Electronics in Agriculture. 2014; 100: 160-167. DOI: 10.1016/j.compag.2013.11.011

[82] Wijethunga P, Samarasinghe S, Kulasiri D, Woodhead I. Digital Image Analysis Based Automated Kiwifruit Counting Technique. 23rd International Conference Image and Vision Computing, New Zealand, Christchurch, Publisher: IEEE, 2008. 1-6. DOI: 10.1109/IVCNZ.2008.4762149

[83] Song Y, Glasbey CA, Horgan GW, Polder G, Dieleman JA, van der Heijdenc GWAM. Automatic fruit recognition and counting from multiple images. Biosystems Engineering. 2014; 118: 203-215. DOI: 10.1016/j.biosystemseng.2013.12.008

[84] Payne AB, Walsh KB, Subedi PP, Jarvis D. Estimation of mango crop yield using image analysis – Segmentation method. Computers and Electronics in Agriculture. 2013; 91: 57-64. DOI: 10.1016/j.compag.2012.11.009

[85] Linker R, Cohen O, Naor A. Determination of the number of green apples in RGB images recorded in orchards. Computers and Electronics in Agriculture. 2012; 81: 45-57. DOI: 10.1016/j.compag.2011.11.007

[86] Narendra VG, Hareesh KS. Quality inspection and grading of agricultural and food products by computer vision—a review. International Journal of Computer Applications. 2010; 2(1): 975-8887. DOI: 10.5120/612-863

[87] Safren O, Alchanatis V, Ostrovsky V, Levi O. Detection of green

apples in hyperspectral images of apple-tree foliage using machine vision. Transactions of the ASABE. 2007; 50(6): 2303-2313. DOI: 10.13031/2013.24083

[88] Nyalala I, Okinda C, Nyalala L, Makange N, Chao Q, Chao L, Yousaf K, Chen K. Tomato volume and mass estimation using computer vision and machine learning algorithms: Cherry tomato model. Journal of Food Engineering. 2019; 263: 288-298. DOI: 10.1016/j.jfoodeng.2019.07.012

[89] Weijun X, Wang F, Yang D. Research on Carrot Grading Based on Machine Vision Feature Parameters. IFAC PapersOnLine. 2019; 52(30): 30-35. DOI: 10.1016/j.ifacol.2019.12.485

[90] Rodriguez-Pulido FJ, Ferrer-Gallego R, Gonzalez-Miret ML, Rivas-Gonzalo JC, Song Y, Glasbey CA, Horgan GW, Polder G, Dieleman JA, van der Heijden GWAM. Automatic fruit recognition and counting from multiple images. Biosystems Engineering. 2014; 118(1), 203-215. DOI: 10.1016/j.biosystemseng.2013.12.008

[91] Aggarwal AK, Mohan R. Aspect ratio analysis using image processing for rice grain quality. International Journal of Food Engineering. 2010; 6(5): 1-14. DOI: 10.2202/1556-3758.1788

[92] Dubosclard P, Larnier S, Konik H, Herbulot A, Devy M. Automated visual grading of grain kernels by machine vision. In: Twelfth International Conference on Quality Control by Artificial Vision. 2015; 9534: 1-8. DOI: 10.1117/12.2182793

[93] Zareiforoush H, Minaei S, Alizadeh MR, Banakar A, Samani BH. Design, development and performance evaluation of an automatic control system for rice whitening machine based on computer vision and fuzzy logic. Computers and Electronics in Agriculture. 2016; 124:14-22. DOI: 10.1016/j.compag.2016.01.024

[94] Bhat S, Panat S, Arunachalam N. Classification of rice grain varieties arranged in scattered and heap fashion using image processing. In: Ninth International Conference on Machine Vision (ICMV 2016). 2017; 10341: 1-6. DOI: 10.1117/12.2268802

[95] Kiliç K, Boyaci IH, Köksel H, Küsmenoglu I. A classification system for beans using computer vision system and artificial neural networks. Journal of Food Engineering. 2007; 78(3): 897-904. DOI: 10.1016/j.jfoodeng.2005.11.030

[96] Venora G, Grillo O, Ravalli C, Cremonini R. Identification of Italian landraces of bean (Phaseolus vulgaris L.) using an image analysis system. Scientia Horticulturae. 2009; 121(4): 410-418. DOI: 10.1016/j.scienta.2009.03.014

[97] Laurent B, Ousman B, Dzudie T, Carl MFM, Emmanuel T. Digital camera images processing of hard-to-cook beans. Journal of Engineering and Technology Research. 2010; 2(9): 177-188. DOI: 10.5897/JETR.9000027

[98] Araújo SA, Alves WAL, Belan PA, Anselmo KP. A computer vision system for automatic classification of most consumed Brazilian beans. In: Bebis G. et al. (eds) Advances in Visual Computing. ISVC 2015. Lecture Notes in Computer Science, Springer, 2015. p. 45-53. DOI: 10.1007/978-3-319-27863-6_5

[99] Belan PA, De Macedo RAG, Pereira MA, Alves WAL, Araújo SA. A fast and robust approach for touching grains segmentation. In: Campilho A., Karray F., ter Haar Romeny B. (eds) Image Analysis and Recognition. Lecture Notes in Computer Science, vol 10882. Springer; 2018. p. 482-489. DOI: 10.1007/978-3-319-93000-8_54

[100] Belan PA, de Macedo RAG, Alves WAL, Santana JCC, Araújo SA. Machine vision system for quality

inspection of beans. The International Journal of Advanced Manufacturing Technology. 2020; 111 (11-12), 1-15. DOI: 10.1007/s00170-020-06226-5

[101] Girolami A, Napolitano F, Faraone D, Braghieri A. Measurement of meat color using a computer vision system. Meat science. 2013; 93(1): 111-118. DOI: 10.1016/j.meatsci.2012.08.010

[102] Jackman P, Sun DW, Allen P. Recent advances in the use of computer vision technology in the quality assessment of fresh meats. Trends in Food Science and Technology. 2011; 22 (4), 185-197. DOI: 10.1016/j.tifs.2011.01.008

[103] Chen K, Sun X, Qin C, Tang X. Color grading of beef fat by using computer vision and support vector machine. Computers and Electronics in Agriculture. 2010; 70(1): 27-32. DOI: 10.1016/j.compag.2009.08.006

[104] Mancini RA, Hunt MC. Current research in meat color. Meat Sci. 2005; 71(1):100-121. DOI: 10.1016/j.meatsci.2005.03.003

[105] Zheng C, Sun DW, Zheng L. Correlating color to moisture content of large cooked beef joints by computer vision. Journal of Food Engineering. 2006; 77(4): 858-863. DOI: 10.1016/j.jfoodeng.2005.08.013

[106] Teimouri N, Omid M, Mollazade K, Mousazadeh H, Alimardani R, Karstoft H. On-line separation and sorting of chicken portions using a robust vision-based intelligent modeling approach. Biosystems Engineering. 2018; 167(4): 8-20. DOI: 10.1016/j.biosystemseng.2017.12.009

[107] Gu J, He N, Wu X. A New Detection Method for Fish Freshness. In Computational Intelligence and Design (ISCID), 2014 Seventh International Symposium 2. 2014; 555-558. DOI: 10.1109/ISCID.2014.153

[108] Vatansever F, Hamblin MR. Far infrared radiation (FIR): its biological effects and medical applications, Photonics Lasers Med. 2012; 1(4): 255-266. DOI: 10.1515/plm-2012-0034

[109] Jain RK, Hoffman AJ, Jepsen PU, LIU PQ, Turchinovich D, Vitiello MS. Mid-infrared, long-wave infrared, and terahertz photonics: introduction. Optics Express. 2020; 28(9): 14169-14175. DOI: 10.1364/OE.395165

[110] Bae TW. Spatial and temporal bilateral filter for infrared small target enhancement. Infrared Physics and Technology. 2014; 63: 42-53. doi.org/10.1016/j.infrared.2013.12.007

[111] Kim DK. Flame detection using region expansions and on-line variances in infrared image. Journal of Korea Multimedia Society. 2009; 12(11): 1547-1556.

[112] Huang H, Yu H, Xu H, Ying Y. Near infrared spectroscopy for on/in-line monitoring of quality in foods and beverages: A review. Journal of Food Engineering. 2008; 87(3): 303-313. DOI: 10.1016/j.jfoodeng.2007.12.022

[113] Peng Y, Wenxiu W. Application of near-infrared spectroscopy for assessing meat quality and safety. In: Theophanides T, editors. Handbook of Infrared Spectroscopy - Anharmonicity of Biomolecules, Crosslinking of Biopolymers, Food Quality and Medical Applications. Ist ed. IntechOpen; 2015. p. 137-163. DOI: 10.5772/58912

[114] Scotter C. Use of near infrared spectroscopy in the food industry with particular reference to its applications to on/in-line food processes. Food Control. 1990; 1(3): 142-149. DOI: 10.1016/0956-7135(90)90006-X

[115] Workman J, Weyer L. Practical guide and spectral atlas for interpretive near-infrared spectroscopy. 2nd ed. CRC Press, Taylor & Francis Group; 2012. 326 p. DOI: 10.1201/b11894

[116] Aenugu HPR, Kumar DS, Parthiban SN, Ghosh S, Banji D. Near Infra-Red Spectroscopy—An Overview. International Journal of ChemTech Research. 2011; 2011, 3: 825-836. ISSN: 0974-4290

[117] Genot V, Bock L, Dardenne P, Colinet G. Use of near-infrared reflectance spectroscopy in soil analysis. A review. Biotechnology, Agronomy, Society and Environment. 2014; 18: 247-261.

[118] Liu Y, Chen YR, Kim MS, Chan DE, Lefcourt AM. Development of simple algorithms for the detection of fecal contaminants on apples from visible/near infrared hyperspectral reflectance imaging. Journal of Food Engineering. 2007; 81(2): 412-418. DOI: 10.1016/j.jfoodeng.2006.11.018

[119] Park B, Lawrence KC, Windham WR, Smith D. Performance of hyperspectral imaging system for poultry surface fecal contaminant detection. Journal of Food Engineering. 2006; 75(3): 340-348. DOI: 10.1016/j.jfoodeng.2005.03.060

[120] Ariana D, Lu R, Guyer D.E. Hyperspectral reflectance imaging for detection of bruises on pickling cucumbers. Computers and Electronics in Agriculture. 2006; 53(1): 60-70. DOI: 10.1016/j.compag.2006.04.001

[121] Xiong Z, Sun DW, Xie A, Han Z, Wang L. Potential of hyperspectral imaging for rapid prediction of hydroxyproline content in chicken meat. Food Chemistry. 2015; 175: 417-422. DOI: 10.1016/j.foodchem.2014.11.161

[122] Gowen AA, O'Donnell CP, Cullen PJ, Downey G, Frias JM. Hyperspectral imaging e an emerging process analytical tool for food quality and safety control. Trends in Food Science and Technology. 2007; 18(2): 590-598. DOI: 10.1016/j.tifs.2007.06.001

[123] Monteiro S, Minekawa Y, Kosugi Y, Akazawa T, Oda K. Prediction of sweetness and amino acid content in soybean crops from hyperspectral imagery. ISPRS Journal of Photogrammetry and Remote Sensing. 2007; 62(1): 2-12. DOI: 10.1016/j.isprsjprs.2006.12.002

[124] Smail V, Fritz A, Wetzel D. Chemical imaging of intact seeds with NIR focal plane array assists plant breeding. Vibrational Spectroscopy. 2006; 42(2): 215-221. DOI: 10.1016/j.vibspec.2006.02.004

[125] Uno Y, Prasher S, Lacroix R, Goel P, Karimi Y, Viau A, Patel R.M. Artificial neural networks to predict corn yield from Compact Airborne Spectrographic Imager data. Computers and Electronics in Agriculture. 2005; 47 (2): 149-161. DOI: 10.1016/j.compag.2004.11.014

[126] Lewis E, Schoppelrei J, Lee E, Kidder L. Near-infrared chemical imaging as a process analytical tool. In: Bakeev K, editor, Handbook of Process analytical technology. 1st ed. Oxford: Blackwell Publishing; 2005. p. 187-225. DOI: 10.1002/9780470988459.ch7

[127] Koehler F, Lee E, Kidder L, Lewis N. Near infrared spectroscopy: the practical chemical imaging solution. Spectroscopy Europe. 2002; 14(3): 12-19.

[128] Mahesh S, Manickavasagan A, Jayas DS, Paliwal J, White, NDG. Feasibility of near-infrared hyperspectral imaging to differentiate Canadian wheat classes. Biosystems Engineering. 2008; 101(1): 50-57. DOI: 10.1016/j.biosystemseng.2008.05.017

[129] Singh CB, Jayas DS, Paliwal J, White NDG. Identification of insect-damaged wheat kernels using short-wave nearinfrared hyperspectral and digital colour imaging. Computers and Electronics in Agriculture. 2010; 73(2): 118-125. DOI: 10.1016/j.compag.2010.06.001

[130] Ariana DP, Lu R. Quality evaluation of pickling cucumbers using hyperspectral reflectance and transmittance imaging: Part II. Performance of a prototype. Sensing and Instrumentation for Food Quality and Safety. 2008; 2(3): 152-160. DOI: 10.1007/s11694-008-9058-9

[131] Shaw G, Manolakis D. Signal processing for hyperspectral image exploitation. IEEE Signal Processing Magazine. 2002; 19(1): 12-16. DOI: 10.1109/79.974715

[132] Lu RF, Chen YR. Hyperspectral imaging for safety inspection of food and agricultural products. In: SPIE Conference on Pathogen Detection and Remediation for Safe Eating, Boston. 1998; DOI: 10.1117/12.335771

[133] Siche R, Vejarano R, Aredo V, Velasquez L, Saldana E, Quevedo R. Evaluation of Food Quality and Safety with Hyperspectral Imaging (HSI). Food Engineering Reviews. 2016; 8(3): 306-322. DOI: 10.1007/s12393-015-9137-8

[134] Cheng JH, Sun D.W. Hyperspectral imaging as an effective tool for quality analysis and control of fish and other seafoods: Current research and potential applications. Trends in Food Science and Technology. 2014; 37(2): 78-91. DOI: 10.1016/j.tifs.2014.03.006

[135] Huang H, Liu L, Ngadi MO. Recent developments in hyperspectral imaging for assessment of food quality and safety. Sensors. 2014; 14(4): 7248-7276. DOI: 10.3390/s140407248

[136] Dumont J, Hirvonen T, Heikkinen V, Mistretta M, Granlund L, Himanen K, Fauch L, Porali I, Hiltunen J, Keski-Saari S, Nygren M, Oksanen E, Hauta-Kasari M, Keinänen M. Thermal and hyperspectral imaging for Norway spruce (Picea abies) seeds screening. Computers and Electronics in Agriculture. 2015; 116: 118-24. DOI: 10.1016/j.compag.2015.06.010

[137] Manley M, Mcgoverin CM, Engelbrecht P, Geladi P. Infuence of grain topography on near infrared hyperspectral images. Talanta. 2012; 89 (2): 223-230. DOI: 10.1016/j.talanta.2011.11.086

[138] Ambrose A, Kandpal LM, Kim MS, Lee WH, Cho BK. High speed measurement of corn seed viability using hyperspectral imaging. Infrared Physics and Technology. 2016; 75: 173-179. DOI: 10.1016/j.infrared.2015.12.008

[139] McGoverin CM, Engelbrecht P, Geladi P, Manley M. Characterisation of non-viable whole barley, wheat and sorghum grains using nearinfrared hyperspectral data and chemometrics. Analytical and Bioanalytical Chemistry. 2011; 401(7): 2283-2289. DOI: 10.1007/s00216-011-5291-x

[140] Polak A, Coutts FK, Murray P, Marshall S. Use of hyperspectral imaging for cake moisture and hardness prediction. IET Image Processing. 2019; 13(7): 1152-1160. DOI: 10.1049/iet-ipr.2018.5106

[141] Chen H, Qiao H, Feng Q, Xu L, Lin Q, Cai K. Rapid Detection of Pomelo Fruit Quality Using Near-Infrared Hyperspectral Imaging Combined With Chemometric Methods. Frontiers in Bioengineering and Biotechnology. 2021; 8: 616943. DOI: 10.3389/fbioe.2020.616943

[142] Gao Z, Shao Y, Xuan G, Wang Y, Liu Y, Han X. Real-time hyperspectral imaging for the in-field estimation of strawberry ripeness with deep learning. Artificial Intelligence in Agriculture. 2020; 4: 31-38. DOI: 10.1016/j.aiia.2020.04.003

[143] Pu YY, Sun DW, Buccheri M, Grassi M, Cattaneo TMP, Gowen A. Ripeness classification of Bananito fruit (Musa acuminata, AA): a comparison study of visible spectroscopy and hyperspectral imaging. Food Anal. Methods. 2019; 12: 1693-1704. doi.org/10.1007/s12161-019-01506-7

[144] González-Cabrera M, Domínguez-Vidal A, Ayora-Cañada M.J. Hyperspectral FTIR imaging of olive fruit for understanding ripening processes. Postharvest Biology and Technology. 2018; 145: 74-82. DOI: 10.1016/j.postharvbio.2018.06.008

[145] Munera S, Amigo JM, Blasco J, Cubero S, Talens P, Aleixos N. Ripeness monitoring of two cultivars of nectarine using VIS-NIR hyperspectral reflectance imaging. Journal of Food Engineering. 2017; 214: 29-39. DOI: 10.1016/j.jfoodeng.2017.06.031

[146] Zhang C, Guo C, Liu F, Kong W, He Y, Lou B. Hyperspectral imaging analysis for ripeness evaluation of strawberry with support vector machine. Journal of Food Engineering. 2016; 179: 11-18. DOI: 10.1016/j.jfoodeng.2016.01.002

[147] Lu Y, Huang Y, Lu R. Innovative hyperspectral imaging-based techniques for quality evaluation of fruits and vegetables: a review. Applied Sciences 2017; 7(2): 1-36. doi:10.3390/app7020189

[148] Zeng X, Miao Y, Ubaid S, Gao X, Zhuang S. Detection and classification of bruises of pears based on thermal images. Postharvest Biology and Technology. 2020; 161(1): 111090. DOI: 10.1016/j.postharvbio.2019.111090

[149] Zhu X, Lia G. Rapid detection and visualization of slight bruise on apples using hyperspectral imaging. International journal of food properties. 2019; 22(1): 1709-1719. DOI: 10.1080/10942912.2019.1669638

[150] Fan S, Li C, Huang W, Chen L. Detection of blueberry internal bruising over time using NIR hyperspectral reflectance imaging with optimum wavelengths. Postharvest Biology and Technology. 2017; 134: 55-66. DOI: 10.1016/j.postharvbio.2017.08.012

[151] Zhang BH, Li JB, Fan SX, Huang WQ, Zhao CJ, Liu CL, Huang D. F. Hyperspectral Imaging Combined with Multivariate Analysis and Band Math for Detection of Common Defects on Peaches (Prunus Persica). Computers and Electronics in Agriculture. 2015; 114: 14-24. DOI: 10.1016/j.compag.2015.03.015

[152] Qiang L, Mingjie T. Detection of Hidden Bruise on Kiwi fruit Using Hyperspectral Imaging and Parallelepiped Classification. Procedia Environmental Sciences. 2012; 12: 1172-1179. DOI: 10.1016/j.proenv.2012.01.404

[153] Li J, Rao X, Ying Y. Detection of common defects on oranges using hyperspectral reflectance imaging. Computers and Electronics in Agriculture. 2011; 78(1): 38-48. DOI: 10.1016/j.compag.2011.05.010

[154] Xing J, Baerdemaeker JD. Bruise detection on 'Jonagold' apples using hyperspectral imaging. Postharvest Biology and Technology. 2005; 37(2): 152-162. DOI: 10.1016/j.postharvbio.2005.02.015

[155] Mehl PM, Chen YR, Kim MS, Chan DE. Development of hyperspectral imaging technique for the detection of

apple surface defects and contaminations. Journal of Food Engineering. 2004; 61(1): 67-81. DOI: 10.1016/S0260-8774(03)00188-2

[156] Kandpal LM, Cho BK. A review of the applications of spectroscopy for the detection of microbial contaminations and defects in agro foods. Journal of Biosystems Engineering. 2014; 39(3): 215-226. DOI: 10.5307/JBE.2014.39.3.215

[157] Kim G, Kim GH, Park J, Kim DY, Cho BK. Application of infrared lock-in thermography for the quantitative evaluation of bruises on pears. Infrared Physics and Technology. 2014; 63: 133-139. DOI: 10.1016/j.infrared.2013.12.015

[158] Lu R, Chen YR, Park B, Choi K.H. Hyperspectral imaging for detecting bruises in apples. In: ASAE Annual International Meeting, Paper No. 993120, St. Joseph, Michigan. 1999.

[159] Lu R. Detection of bruise on apples using near-infrared hyperspectral imaging. Transactions of the ASAE. 2003; 46(2): 1-8. DOI: 10.13031/2013.12941

[160] Changyeun M, Giyoung K, Moon SK, Jongguk L, Kangjin L, Wang-Hee L, Byoung-Kwan C. On-line fresh-cut lettuce quality measurement system using hyperspectral imaging. Biosystems engineering. 2017; 156: 38-50. DOI: 10.1016/j.biosystemseng.2017.01.005

[161] Yang B, Gao Y, Yan Q, Qi L, Zhu Y, Wang B. Estimation Method of Soluble Solid Content in Peach Based on Deep Features of Hyperspectral Imagery. Sensors. 2020; 20(18): 1-12. DOI: 10.3390/s20185021

[162] Pu Y, Sun D, Riccioli C, Buccheri M, Grassi M, Cattaneo TM, Gowen A. Calibration transfer from micro nir spectrometer to hyperspectral imaging: A case study on predicting soluble solids content of bananito fruit (Musa acuminata). Food Analytical Methods. 2018; 11(11): 1021-1033. DOI: 10.1007/s12161-017-1055-3

[163] Ma T, Li X, Inagaki T, Yang H, Tsuchikawa S. Noncontact evaluation of soluble solids content in apples by near-infrared hyperspectral imaging. Journal of Food Engineering. 2017; 224: 53-61. DOI: 10.1016/j.jfoodeng.2017.12.028

[164] Pu H, Liu D, Wang L, Sun DW. Soluble Solids Content and pH Prediction and Maturity Discrimination of Lychee Fruits Using Visible and Near-Infrared Hyperspectral Imaging. Food Analytical Methods. 2016; 9(1): 235-244. DOI: 10.1007/s12161-015-0186-7

[165] Li J, Peng YK, Chen LP, Huang WQ. Near-infrared hyperspectral imaging combined with cars algorithm to quantitatively determine soluble solids content in "Ya" pear. Spectroscopy and Spectral Analysis. 2014; 34(5): 1264-1269. DOI: 10.3964/j.issn.1000-0593(2014)05-1264-06

[166] Leiva-Valenzuela GA, Lu R, Aguilera JM. Prediction of firmness and soluble solids content of blueberries using hyperspectral reflectance imaging. 2013; Journal of Food Engineering. 115 (1): 91-98. DOI: 10.1016/j.jfoodeng.2012.10.001

[167] Baiano A, Terracone C, Peri G, Romaniello R. Application of hyperspectral imaging for prediction of physico-chemical and sensory characteristics of table grapes. Computers and Electronics in Agriculture. 2012; 87: 142-151. DOI: 10.1016/j.compag.2012.06.002

[168] Rajkumar P, Wang N, EImasry G, Raghavan GSV, Gariepy Y. Studies on banana fruit quality and maturity stages using hyperspectral imaging. Journal of Food Engineering. 2012; 108(1):

194-200. DOI: 10.1016/j.jfoodeng.2011.05.002

[169] Mendoza F, Lu R, Ariana D, Cen H, Bailey B. Integrated spectral and image analysis of hyperspectral scattering data for prediction of apple fruit firmness and soluble solids content. Postharvest Biology and Technology. 2011; 62(2): 149-160. DOI: 10.1016/j.postharvbio.2011.05.009

[170] Peng Y, Lu R. Analysis of spatially resolved hyperspectral scattering images for assessing apple fruit firmness and soluble solids content. Postharvest Biology and Technology. 2008; 48(1): 52-62. DOI: 10.1016/j.postharvbio.2007.09.019

[171] ElMasry G, Wang N, ElSayed A, Ngadi M. Hyperspectral imaging for nondestructive determination of some quality attributes for strawberry. Journal of Food Engineering. 2007; 81 (1): 98-107. DOI: 10.1016/j.jfoodeng.2006.10.016

[172] El Fakir C, Poffo L, Billiot B, Besnard P, Goujon JM. Active Hyperspectral Mid-Infrared Imaging Based on Widely Tunable QCL Laser. Conference: 21st International Conference on Transparent Optical Networks (ICTON), Angers, France. 2019; 1-4. DOI: 10.1109/ICTON.2019.8840448

[173] Buitrago MF, Groen TA, Hecker CA, Skidmore AK. Spectroscopic determination of leaf traits using infrared spectra. International journal of applied earth observation and geoinformation. 2018; 69: 237-250. DOI: 10.1016/j.jag.2017.11.014

[174] Ribeiro da Luz B, Crowley JK. Spectral reflectance and emissivity features of broad leaf plants: Prospects for remote sensing in the thermal infrared (8.0-14.0 μm). Remote Sensing of Environment. 2007; 109(4): 393-405. DOI: 10.1016/j.rse.2007.01.008

[175] Ribeiro da Luz B. Attenuated total reflectance spectroscopy of plant leaves: a tool for ecological and botanical studies. New phytologist. 2006; 172(2): 305-318. DOI: 10.1111/j.1469-8137.2006.01823.x

[176] Chrzanowski K. Testing thermal imagers: Practical guidebook. Warsaw: Military University of Technology. 2010; INSB 978-83-61486-81-7

[177] Minkina W, Dudzik S. Handbook of Infrared thermography: Errors and uncertainties. 1st ed. Chichester: Wiley; 2009. 212 p. ISBN-10: 0470747188

[178] Ball M, Pinkerton H. Factors affecting the accuracy of thermal imaging cameras in volcanology. Journal of Geophysical Research Atmospheres. 2006; 111(B11): 1-14. DOI: 10.1029/2005JB003829

[179] Chelladurai V, Jayas DS, White NDG. Thermal imaging for detecting fungal infection in stored wheat. Journal of Stored Products Research. 2010; 46(3): 174-179. DOI: 10.1016/j.jspr.2010.04.002

[180] Baranowski P, Mazurek W, Witkowska-Walczak B, Sławiński C. Detection of early apple bruises using pulsed-phase thermography. Postharvest Biology and Technology. 2009; 53(3): 91-100. DOI: 10.1016/j.postharvbio.2009.04.006

[181] Baranowski P, Mazurek W, Wozniak J, Majewska U. Detection of early bruises in apples using hyperspectral data and thermal imaging. Journal of Food Engineering. 2012; 110 (3): 345-355. DOI: 10.1016/j.jfoodeng.2011.12.038

[182] Kuzy J, Jiang Y, Li C. Blueberry bruise detection by pulsed thermographic imaging. Postharvest

Biology and Technology. 2018; 136: 166-177. DOI: 10.1016/j.postharvbio.2017.10.011

[183] Takacs R, Jovicic V, Rasic AZ, Geier D, Delgado A, Becker T. Evaluation of baking performance by means of mid-infrared imaging. Innovative Food Science and Emerging Technologies. 2020; 61: 102327. DOI: 10.1016/j.ifset.2020.102327

[184] Lopez-Sanchez J, Fortuny-Guasch J. 3-D radar imaging using range migration techniques. IEEE Transactions on Antennas and Propagation. 2000; 48(5): 728-737. DOI: 10.1109/8.855491

[185] Sheen D, McMakin D, Hall T. Three-dimensional millimeter-wave imaging for concealed weapon detection. IEEE Transactions on Microwave Theory and Techniques. 2001; 49(9): 1581-1592. DOI: 10.1109/22.942570

[186] Cook BS, Tehrani B, Cooper JR, Tentzeris MM. Multilayer inkjet printing of millimeter-wave proximity-fed patch arrays on flexible substrates. IEEE Antennas and Wireless Propagation Letters. 2013; 12: 1351-1354. DOI: 10.1109/LAWP.2013.2286003

[187] Scott S. Three-Dimensional Microwave Imaging for Indoor Environments [Ph.D.]. University of California; 2017.

[188] Fear EC, Meaney PM, Stuchly MA. Microwaves for breast cancer detection? IEEE Potentials. 2003; 22(1): 12-18. DOI: 10.1109/MP.2003.1180933

[189] Fear EC, Hagness SC, Meaney PM, Okonieweski M, Stuchluy MA. Enhancing breast tumor detection with near-field imaging. IEEE Microwave Magazine. 2002; 3(1): 48-56. DOI: 10.1109/6668.990683

[190] Ambrosanio M, Scapaticci R, Crocco L. A Simple Quantitative Inversion Approach for Microwave Imaging in Embedded Systems. International Journal of Antennas and Propagation. 2015; 12: 1-18. DOI: 10.1155/2015/129823

[191] Manickavasagan A, Jayasuriya H, editors. Handbook of Imaging with Electromagnetic Spectrum. 1st ed. (eds.). Berlin, Heidelberg: Springer; 2014. 204 p. DOI: 10.1007/978-3-642-54888-8

[192] Pastorino M, editor. Handbook of Microwave Imaging. 1st ed. John Wiley & Sons; 2010. 304 p. ISBN: 978-0-470-60247-8

[193] Donelli M, Franceschini D, Rocca P, Massa A. Three-dimensional microwave imaging problems solved through an efficient multiscaling particle swarm optimization. IEEE Transactions on Geoscience and Remote Sensing. 2009; 47(5):1467-1481. DOI: 10.1109/TGRS.2008.2005529

[194] Benedetti M, Donelli M, Lesselier D, Massa A. A two-step inverse scattering procedure for the qualitative imaging of homogeneous cracks in known host media-preliminary results. IEEE Antennas and Wireless Propagation Letters. 2007; 6: 592 - 595. DOI: 10.1109/LAWP.2007.910954

[195] Huang T, Mohan AS. A microparticle swarm optimizer for the reconstruction of microwave images. IEEE Transactions on Antennas and Propagation. 2007; 55(3 I): 568-576. DOI: 10.1109/TAP.2007.891545

[196] Donelli M, Massa A. Computational approach based on a particle swarm optimizer for microwave imaging of two-dimensional dielectric scatterers. IEEE Transactions on Microwave Theory and Techniques. 2005; 53(5), 1761-1776. DOI: 10.1109/TMTT.2005.847068

[197] Bort E, Donelli M, Martini A, Massa A. An adaptive weighting

strategy for microwave imaging problems. IEEE Transactions on Antennas and Propagation. 2005; 53(5): 1858-1862. DOI: 10.1109/TAP.2005.846811

[198] Caorsi S, Massa A, Pastorino M, Donelli M. Improved microwave imaging procedure for non-destructive evaluations of two-dimensional structures. IEEE Transactions on Antennas and Propagation. 2004; 52(6): 1386-1397. DOI: 10.1109/TAP.2004.830254

[199] Giordano A. Microwave Imaging Technology for Food Contamination Monitoring [MSc]. College of Electronic Engineering, Telecommunications and Physics, Politecnico Di Torino; 2018.

[200] Ragni L, Al-Shami A, Mikhaylenko G, Tang J. Dielectric characterization of hen eggs during storage. Journal of Food Engineering. 2007; 82(4): 450-459. DOI: 10.1016/j.jfoodeng.2007.02.063

[201] Guo W, Trabelsi S, Nelson SO, Jones DR. Storage effects on dielectric properties of eggs from 10 to 1800 MHz. Journal of Food Science. 2007; 72: 335-340. DOI: 10.1111/j.1750-3841.2007.00392.x

[202] Tai TC, Wu HW, Hung CY, Wang YH. Food Security Sensing System Using a Waveguide Antenna Microwave Imaging through an Example of an Egg. Sensors. 2019; 20 (3): 699. DOI:10.3390/s20030699.

[203] Abdullah MZ, Guan LC, Lim KC, Karim AA. The applications of computer vision system and tomographic radar imaging for assessing physical properties of food. Journal of Food Engineering. 2004; 61(1): 125-135. DOI: 10.1016/S0260-8774(03)00194-8

[204] Sándor K, Zsolt P, Ádám C, György B, Tamás M, Tamás D. Non-destructive imaging and spectroscopic techniques to investigate the hidden-lifestyle arthropod pests: a review. Journal of Plant Diseases and Protection. 2020; 127: 283-295. doi.org/10.1007/s41348-020-00300-6

[205] Singh V, Sharma N, Singh S. A review of imaging techniques for plant disease detection. Artificial Intelligence in Agriculture. 2020; 4: 229-242. DOI: 10.1016/j.aiia.2020.10.002

[206] Hart AG, Bowtell RW, Köfckenberger W, Wenseleers T, Ratnieks FLW. Magnetic resonance imaging in entomology: a critical review. Journal of Insect Science. 2003; 3(5),1-9. DOI: 10.1673/031.003.0501

[207] Joyce DC, Hockings PD, Mazucco RA, Shorter AJ. 1H-Nuclear magnetic resonance imaging of ripening 'Kensington Pride' mango fruit. Functional Plant Biology. 2002; 29(7): 873-879. DOI: 10.1071/PP01150

[208] Thybo AK, Szczypinski PM, Karlsson AH, Dønstrup S, Stødkilde-Jørgensen HS, Andersen HJ. Prediction of sensory texture quality attributes of cooked potatoes by NMR-imaging (MRI) of raw potatoes in combination with different image analysis methods. Journal of Food Engineering. 2004; 61 (1): 91-100. DOI: 10.1016/S0260-8774 (03)00190-0

[209] Koizumi M, Naito S, Ishida N, Haishi T, Kano H. A dedicated MRI for food science and agriculture. Food Science and Technology Research. 2008; 14 (1): 74 - 82. DOI: 10.3136/fstr.14.74

[210] Taglienti A, Massantini R, Botondi R, Mencarelli F, Valentini M. Postharvest structural changes of Hayward kiwifruit by means of magnetic resonance imaging spectroscopy. Food Chemistry. 2009; 114(4): 1583-1589. DOI: 10.1016/j.foodchem.2008.11.066

[211] Ciampa A, Dell'Abate MT, Masetti O, Valentini M, Sequi P.

Seasonal chemical-physical changes of PGI Pachino cherry tomatoes detected by magnetic resonance imaging. Food Chemistry. 2010; 122(4): 1253-1260. DOI: 10.1016/j.foodchem.2010.03.078

[212] Zhang L, McCarthy MJ. Measurement and evaluation of tomato maturity using magnetic resonance imaging. Postharvest Biology and Technology. 2012; 67, 37-43. DOI: 10.1016/j.postharvbio.2011.12.004

[213] Melado-Herrerosa A, Munoz-Garcíab MA, Blancoc A, Valc J, Fernández-Valled M, Barreiro P. Assessment of watercore development in apples with MRI: Effect of fruit location in the canopy. Postharvest Biology and Technology. 2013; 86: 125-133. DOI: 10.1016/j.postharvbio.2013.06.030

[214] Divya S, Thyagarajan D, Sujatha G. Magnetic resonance imaging technology for process control and quality maintenance in food quality operation. International Journal of Engineering and Technology (IJET). 2013; 4(6): 441-449. ISSN: 0975-4024

[215] Van As H, Van Duynhovenc J. MRI of plants and foods. Journal of Magnetic Resonance. 2013; 229: 25-34. DOI: 10.1016/j.jmr.2012.12.019

[216] Chizhik VI, Chernyshev YS, Donets AV, Frolov VV, Komolkin AV, Shelyapina MG. Basic principles of detection of nuclear magnetic resonance. In: Chizhik VI, Chernyshev YS, Donets AV, Frolov VV, Komolkin AV, Shelyapina MG. Handbook of Magnetic resonance and its applications. 1st ed. Chizhik, Cham: Springer; 2014. P. 93-162. DOI: 10.1007/978-3-319-05299-1_2

[217] Herremansa E, Melado-Herrerosb A, Defraeyea T, Verlindenc B, Hertoga M, Verbovena P, Vald J, Fernández-Vallee ME, Bongaersf E, Estradeg P, Weversh M, Barreirob P, Nicolaïa BM. Comparison of X-ray CT and MRI of watercore disorder of different apple cultivars. Postharvest Biology and Technology. 2014; 87: 42-50. DOI: 10.1016/j.postharvbio.2013.08.008

[218] Patel KK, Khan MA, Kar A. Recent developments in applications of MRI techniques for foods and agricultural produce—An overview. Journal of Food Science and Technology. 2015; 52(1): 1-26. DOI: 10.1007/s13197-012-0917-3

[219] Pfugfelder D, Metzner R, van Dusschoten D, Reichel R, Jahnke S, Koller R. Non-invasive imaging of plant roots in diferent soils using magnetic resonance imaging (MRI). Plant Methods. 2017; 13(102): 1-9. DOI: 10.1186/s13007-017-0252-9

[220] Kamal T, Cheng S, Khan IA, Nawab K, Zhang T, Song Y, Wang S, Nadeem M, Riaz M, Khan MAU, Zhu BW, Tan M. Potential uses of LF-NMR and MRI in the study of water dynamics and quality measurement of fruits and vegetables. Journal of Food Processing and Preservation. 2019; 43 (474): 1-21. DOI: 10.1111/jfpp.14202

[221] Baek S, Lim J, Lee JG, McCarthy MJ, Kim SM. Investigation of the maturity changes of cherry tomato using magnetic resonance imaging. Applied Sciences. 2020; 10(15): 5188. DOI: 10.3390/app10155188